# scorched

## South Africa's changing climate

# scorched

## South Africa's changing climate

Leonie S Joubert

WITS UNIVERSITY PRESS

Wits University Press
1 Jan Smuts Avenue
Johannesburg
http://witspress.wits.ac.za

© 2006 Leonie S Joubert

First printed 2006
Second printing 2007

ISBN 978-1-86814-437-2



Copyediting by Judith Marsden
Cover design and typesetting by Crazy Cat Designs
Printing and binding by Creda Communications

For Wayne

who cares about the things that matter

What a piece of work is man!
How noble in reason! how infinite in faculties!
In form and moving, how express and admirable!
In action how like an angel!
In apprehension, how like a god!
The beauty of the world! the paragon of animals!
And yet, to me, what is this quintessence of dust?

**William Shakespeare, *Hamlet***

# Contents

# Acknowledgements

This book started with an unobtrusive advertisement tucked away in the depths of an industry newsletter:

*Opportunity to collaborate with climate change research group. Candidate must be enrolled in a Masters degree in Science Journalism at Stellenbosch University. Willingness to travel to Marion Island a plus.*

Antarctic travel is embedded in the fabric of my clan and I have often felt the tug of that frozen wilderness. I dug out my atlas and did the maths. Marion Island was about as close to half way as I was likely to get in the foreseeable future. I jumped on board the Masters programme faster than you can shake a Patagonian toothfish at a king penguin. The detour which followed was dramatic enough to divert my attention from Antarctica to something just as compelling.

For this opportunity, I extend my thanks to Prof. Steven Chown of the Capacity Building Programme for Climate Change Research (CBP-CCR) and director of the DST-NRF Centre of Excellence for Invasion Biology at Stellenbosch University. His foresight brought a journalist onto his team and included the kind of funding that allows a writer the luxury of further study. The CBP-CCR put me through a Masters degree, subsidised my expenses and floated the cost of research. This book wouldn't have become more than a few pen scratches on paper without that vision.

Behind the CBP-CCR was the considerable support of US AID, Stellenbosch University and the South African Department of Environmental Affairs and Tourism. The CBP-CCR's Richard Mercer and Jacques Deere offered logistical support with unfailing good humour.

Thanks to all the scientists who opened the bowels of their work to my clumsy poking about and patiently fielded the repeated onslaught of questions: Prof. Steven Chown, Prof. Melodie McGeoch, Dr Guy Midgley, Prof. Bruce Hewitson, Prof. William Bond, Dr Bob Scholes, Prof. Norman Owen-Smith, Prof. Marthan Bester, John Cooper, Dr Case Rijsdijk, Atherton de Villiers, Andrew Turner, Dr Les Minter, Dr André Claassens, Alan Heath, Dr Brian Henen, Dr Thomas Leuteritz, Prof. Les Underhill, Doug Harebottle, Prof. Michael Schleyer, Wendy Foden, Ernst van Jaarsveld, Dr Larry Hutchings, Lara Atkinson, Dr Scott Drimie, Dr Barend Erasmus, Prof. Roland Schulze, Dr Anton Pauw, the Botanical Society's director of conservation Mark Botha, Harald Winkler, and many more who may not have been quoted directly but certainly helped to inform my thinking.

Tony Leiman at UCT's School of Economics kindly talked me through the implications of Kyoto. Prof. William Ruddiman allowed the use of his graphs, and the South African National Biodiversity Institute (SANBI) generously permitted the reproduction of its maps.

There are many more people who shaped things along the way: Prof. Paul Maylam and Dr Julian Cobbing, historians who turned my world view on its head; Steve Bolnick, who showed me elephants up close; James Clarke and Dr Ian Player, giants on whose shoulders the rest of us might stand; Brian Glennon, who dreamed of journalism and played a mean guitar; Brian Garman, who talked about 'pootling' across Petri dishes; Dr Don Pinnock, one of my favourite natural history writers, for hours of conversation on the slopes of Table Mountain; David Bristow, for the books; Cameron Ewart-Smith, who survived the early manuscript and plenty of tea-fuelled philosophising; John Stokes, for allowing me to squat in the Inxcom offices from time to time; Lesley Beake, Kevin Kidson and the rest of the wine.co.za team for the shared vision and regular work; and Veronica Klipp and Melanie Pequeux at Wits University Press who delivered the baby.

Thanks, too, to all my friends, who cheered along the way: Belinda Anderson, Lin Murray and Alv Visser, from the formative years; Kabs Duiker, who kept me smiling through some bleak times and finished three acclaimed novels before he turned thirty; Clifford and Maryke Roberts, who share the passion for life and words; Linda Martindale, who knew I could before I did; Elize Beltrame felt the same; the inimitable ladies in The Boeksisters' Koekclub (or, sometimes, The Koeksisters' Boekclub) – Margie Jordan, Tessa Redman, Josie Carbone and Karen Collett whose veins flow with the same stuff; my MPhil colleagues Gio Gerbi, Helene Uys and Chantal Rutter; those of you who kept debating the ethics of SUVs and First World excess – Steven and Lara Thomas, Rozi and Neil Andrew, Neil Harrison, Charlotte Kilbane and the rest of the crowd: never allow the indignation to cool.

To my family who, in recent years, each in their own way filled in the potholes, pointed out route marks and gave me *padkos* along the way. You made this journey achievable and, indeed, even comfortable: my parents Bruce and Elena Joubert; Simon and Tracey Joubert; Robin Joubert and Nettie van Zyl; Simon and Pat Culshaw. To Florence and Richard Everitt for the considerable logistical support of No 304.

Most of all, thanks to Wayne de Villiers for the 24/7 technical support and for always being the calm at the centre of the storm.

# Introduction

Were you around to remember the 1970s? The seventies were about the rise of Black Consciousness and the death of Steve Biko; the Soweto riots; John Vorster and PW Botha; rural removals; independence for Mozambique and Angola, and *swart gevaar* ('black danger'); the arrival of television and the reign of Springbok Radio; disco and drive-in movies. The decade started with the death of Jimi Hendrix and the reign of Chairman Mao. It ended with the rise of Maggie Thatcher. Leonid Brezhnev returned somewhere in between. It was about 'glam rock' and David Bowie's *Ziggy Stardust*; Nixon and Watergate; the end of the Vietnam War; the start of China's nuclear tests; *Star Wars* but not yet Reagan's Star Wars; Led Zeppelin and The Bee Gees; punk rock and reggae. The pocket calculator but not quite the PC; the death of Elvis; meltdown on Three Mile Island; the defection of the Russian ballet dancer Mikhail Baryshnikov; the kidnapping of hostages from the Olympic Village in Munich; the Terracotta Army unearthed in China. It came after post-sixties free love but before AIDS-anxious eighties love.

You may wonder about the relevance of all of this, but it's about context. Milestones on the timeline help to peg down context in the same way that telephone poles along the roadside give perspective on a long journey. The seventies was jam-packed with milestones. In 1973 Ethiopia plunged into a famine the likes of which the modern world had never before seen splashed over the front pages of newspapers. World food security became the hot topic in an increasingly globalised media network. The world also learned what happens when the oozing black lifeblood of industry is pinched off at the source. During the 1973 Yom Kippur War, OPEC countries reduced oil supplies to Europe and placed an embargo against the United States for its collaboration with Israel during the conflict. Oil prices quadrupled, crude oil supplies fell and, for a time, industrial nations staggered to their knees. British industry had to be reined in to a three-day work week. The US gas stations ran dry and recession spread across the globe.

The first oil crisis, followed by another towards the close of the decade, bore witness to a new era in world politics. It also brought to startling light how fragile our industrial success is; how dependent its survival is on natural resources that lie in the hands of a few; and how there is by no means a limitless supply of the stuff. The 1970s heralded a new world-view. Neil Armstrong and the lads from NASA had spent enough time orbiting Earth and gadding about on the Moon to switch our perspective of life on the planet. Those first images of a solitary blue-and-white-swirled globe like a small island in a sea of vacant space were a stark reminder of how vulnerable and truly alone we are in the greater scheme of things. Suddenly 'not in my back yard' referred to a whole lot more than the bit of lawn you mowed once a week.

Environmental awareness, which had surfaced in the sixties, picked up the pace in the seventies. Discourse between civil society, scientists and government was beginning to rumble with concerns about pollution, over-exploitation of resources, population growth, food security, and an early awareness of environmental degradation. The United Nations was beginning to pick at the scab by hosting global conferences. Some governments even set up the first ministries of the environment, which until then had not even been an afterthought. Greenpeace launched its first protest against nuclear testing by the United States in Alaska. The scientific community unveiled research which proved that chlorofluorocarbons (CFCs) were a threat to the ozone layer. People were becoming seriously worried about the pressure our expanding population was placing on our communal back yard.

Something else happened in the 1970s: the Northern Hemisphere was warming. There had actually been various ups and downs in the trend of boreal temperature in the preceding century but now scientists were beginning to understand what was going on.

Global warming, or at least a suspicion of it, surfaced long before the seventies. The notion that the gases wrapped around the planet could trap the Sun's heat – the greenhouse effect – had been bandied about since 1824 when a French mathematician and physicist, Baron Jean-Baptiste Joseph Fourier suggested the idea. Then, in 1896, the Swede Svante Arrhenius hypothesised that by pumping more carbon dioxide into the atmosphere through burning coal and other fossil fuels, the greenhouse effect could be enhanced. The human species, he asserted, could facilitate global warming. The broader scientific community dismissed the theory, saying that absorption of carbon dioxide by the oceans and plants would balance out any negative effects of burning fossil fuels.

Thermometers have been around since Galileo but accurate temperature measurement only dates back to about 1861. So when people in the 1930s began to notice something strange about the weather, they only had seventy-odd years of records which were not representative of the globe: most came from the Northern Hemisphere, mainly eastern North America and western Europe. Few long-term records originated from the south, where most of the hemisphere is inaccessible Southern Ocean or icy Antarctic desert and the inhabited regions of the south did not have the strong tradition of modern science that was well established in the north.

Many records came from expanding cities where urban living is known to create 'heat islands' in which cars, industry and buildings generate and trap heat in a way that skews temperature recording. Picking through all this data, it became evident that temperature had climbed slowly and steadily through the late 1800s. In 1939 amateur meteorologist Guy Stewart Callendar calculated a general global increase of nearly 0.5°C between 1890 and 1935, matching a steady increase in atmospheric carbon dioxide. His work never achieved more than some scientific curiosity, partly for the same reason that Arrhenius's work was dismissed. But Callendar's theory could not explain the mitigating factor of water vapour: in a warmer world the atmosphere can hold more water vapour, resulting in greater cloud formation which would shield Earth from incoming radiation and balance out the warming effect.

Other scientists were applying themselves to understanding how natural phenomena played out in climate: cycles in ice ages, glacial periods, volcanoes puffing gaseous clouds into the air, the shifting orbital cycles of the Earth, and pulses in the Sun's output. Separating natural climate shifts from those caused by human activity on the planet remains one of the great challenges in climate change research.

By the 1950s, computing technology had come a long way. Suddenly there were massive silicon brains to take over the workload and grind through volumes of numbers generated by meteorology. When physicist Dr Gilbert Plass wasn't working on the Manhattan Project, he spent his spare time pondering on carbon dioxide and its heat-trapping potential in the atmosphere. In 1956 he announced that his calculations showed the potential for a 1.1°C temperature increase per century owing to the burning of fossil fuels and that we could well be a hazard to ourselves. The world started to take notice.

But there was a problem. Against all comprehension of the subject, weather records showed that temperatures in the Northern Hemisphere had been decreasing since the 1940s. This trend lasted until the 1970s and it would later be shown that the cause was related to sunspot activity. Nevertheless, it muddied the waters in an already highly complicated science.

Meanwhile the Cold War was playing out in infinitely more heated squabbles. The world held its breath as tensions around the Cuban Missile Crisis escalated before mercifully dissipating. Communist Bloc and Western Democratic Bloc countries continued to fail to settle their differences, fighting out their issues by proxy on battlegrounds in the Third World: Vietnam, Korea, Angola, Mozambique . . . The threat of nuclear winter was far more frightening than vague and distant notions of altering global climate.

Fortunately the scientific community continued to apply its communal mind to the problem. By the time the 1970s rolled in, it appeared that the cooling trend in the north had reversed and temperatures there were beginning to climb again. Droughts and other extreme weather events, notably the famine in Ethiopia, had the media abuzz with speculation. The oil crises started people thinking about energy issues. A massive hole was found in a damaged ozone layer. The world was ready to start grappling with global climate change and protection of the atmosphere.

The first World Climate Conference was held in Geneva. The year 1981 was declared the warmest on record and the scientific community entered the 1980s by calling on governments to take action. Science now also understood that there were other ways to get around the troublesome issue of only having a few decades of reliable thermometer readings: they looked at proxy data to see how climate and atmospheric gases had changed through time. Gas trapped in ice caps recorded hundreds of thousands of years of fluctuation. Tree rings showed shifts between cool unproductive years and warmer productive years, giving a shorter-term record. Pollen trapped in layers of ancient mud showed how vegetation had shifted its distribution in response to climate over thousands of years. Likewise, by looking at the different species of plankton whose shell remains are trapped in limestone layers on the seabed, scientists could calculate which species thrived at different times and what the temperature must have been for them to have done so. Shell fossils in the seabed have provided temperature records going back a hundred million years.

Each aspect – ice, tree rings, pollen, seashells – provides a different timeline. Juxtaposed, the chronology of the planet's ancient climate emerges from the haze of the past. Carbon dioxide ($CO_2$) from burning fossil fuel was also understood to be just one of the culprits in anthropogenic climate change. In addition, methane ($CH_4$) and nitrous oxide ($N_2O$) from cattle feed lots, rotting paddy fields and landfills, tillage agriculture and the chemical industry were factored in along with deforestation and other changes in land cover to alter the atmospheric makeup. Aerosols and ozone were also found to feed into the system.

The 1980s brought a call for stronger action. The Montreal Protocol showed that it was possible for global action to reduce emissions into the atmosphere by placing a cap on the production of ozone-depleting CFCs. Then 1988 achieved the new and dubious distinction of being the hottest year on record (which would later be surpassed by an even hotter 1998 and 2005), and the United Nations established the Intergovernmental Panel on Climate Change (IPCC). Atmospheric carbon dioxide concentration was finally pegged down at 350 parts per million (ppm), which was up from the pre-industrial level of 280 ppm. A 500 ppm concentration was believed to be 'dangerous'. A strict call was made for a reduction in greenhouse gas emissions but it was understood that the planet was already committed to a small amount of global climate change.

The decade of the 1990s saw large-scale global collaboration among the scientific community through the IPCC. A 1992 meeting in Rio de Janeiro witnessed over a hundred countries signing the United Nations Framework Convention on Climate Change (UNFCCC), in which they acknowledged the climate system as a shared resource. They produced the first global effort to curb global warming. The United States (US) signed the agreement. However, when in 1997 the members met to ratify the Kyoto Protocol – which committed members to set targets for emissions reduction – the US failed to sanction the agreement. Since the US is responsible for nearly a quarter of all global emissions, its lack of commitment to Kyoto remains one of the most contentious issues in the politics of climate change. It will be looked at more closely in the closing pages of this book.

In 2001 the IPCC released the most up-to-date and comprehensive body of work, *Climate Change 2001 – The Scientific Basis*. The consensus, now, is that the planet is committed to dangerous global warming and yet the juggernaut of industrial growth seems unstoppable.

The environmental movement – which grew out of the momentum it gained in the 1970s – has had to take as good as it gives, rolling with the punches from critics. In some respects the movement is represented by clinical science on one side, radical activists on the other, and any number of shades in between. Environmentalists have been labelled scaremongers and New Age doomsday predictors. Some have called environmentalism the new religion for atheists. I suppose there are activists who follow their cause with religious zeal. But the danger of interpreting environmentalism as a religion is this: it reduces the issues to those of a belief system rather than what, in this case, they really are – the findings of observable scientific fact.

Global warming is not a belief system that you choose to adopt if it suits your political view. Understanding climate change is an extremely complicated tangle, which scientists must try to unravel. Predicting where climate change will go in future is even more difficult. More and more meteorological data is fed into increasingly powerful computers whose algorithms are tailored to simulate future climate in a warmer world. These computer models are not going to predict how much snow will fall for the Vancouver Winter Olympics or what the weather will be like for the Soccer World Cup in South Africa in 2010. They will suggest the trends for different parts of the globe – and even the conservative estimates are worrying.

We now know that the Earth has experienced long slow changes in climate as well as abrupt shifts in the past. In the words of the UNFCCC:

> The climate appears to have 'tipping points' that can send it into sharp lurches and rebounds. Although scientists are still analysing what happened during those earlier events, it's clear that an overstressed world with 6.3[1] billion people is a risky place to be carrying out uncontrolled experiments with the climate.

Critics of climate change still exist, but they have retreated to the margins where fringe groups of sceptics are increasingly losing credibility. One opposing argument maintains that the scientific community and environmental activists have bought into global warming because of the financial benefits – it has become a global industry. If you are a scientist or NGO and you want funding, they argue, just work a climate change angle into research and the money will pour in. Media complicity is explained away by their need for headlines that sell. Another group believes that the current warming trend is entirely natural, probably because of increased sunspot activity. It will reverse fairly soon, they reassure us, thus proving the wider scientific community wrong. Others say that rising heat will increase evaporation, thereby producing more clouds which will block out the heat. The system will balance out nicely in the end and all's well that ends well. Some fringe elements support George W Bush's decision not to participate in the Kyoto Protocol's efforts to reduce greenhouse gas emissions because the whole issue is built on shoddy science.

If you delve into the debate, you will encounter the writing of statistician and political scientist Bjørn Lomborg. His controversial book *The Skeptical Environmentalist – Measuring the Real State of the World* is one of the texts that muddies the water for people like you and me who just want to understand what is going on and to make the necessary consumer and voter decisions accordingly. The issue, Lomborg argues, has nothing to do with whether global warming is taking place, but rather by how much it is happening. He maintains the worst case scenario put forward by the IPCC (a 6°C increase in average temperature within the next century), is overblown and that the lower end of its prediction is a more likely increase: 1.5°C. This small amount of warming will benefit the north as its frozen high latitudes become opened up to agriculture and development by the creeping thaw which is already turning the tundra soft and muddy. The wealthy north can then rescue the floundering south by pouring in aid to assist its countries with adapting to the changes that will leave many parts wracked by drought and further poverty. Lomborg digs out reams of literature which support his positive view of things and ignores a massive body of work which does not.

Science is all about dissidence and building good arguments by disproving unsound ones. The problem is that when a community becomes vastly polarised on a subject, the layperson is often caught in the middle of a debate where facts are hurled about like custard pies in a circus ring. Hit the public enough with this kind of treatment and eventually you will render it catatonic with confusion. Take the HIV/AIDS debate in South Africa. A dissident community argued that AIDS is not caused by the Human Immunodeficiency Virus (HIV). Rather, the weakened immune system so prevalent in the developing world is the consequence of poverty and malnutrition. The government sided with the dissidents and the debate between them and mainstream science spilled over into the news media. The squabbling went on for months while the pandemic spread because important decisions were not made about how to treat the disease.

Another case in point is the nuclear debate. Pro- and anti-nuclear factions each have their basket of scientific facts to throw at each other whenever newspapers give them a column inch. I am not afraid to admit I find it so bewildering that I don't know where to hang my hat on this issue, even though I've written enough of my own column inches on the subject.

On the Lomborg-versus-IPCC debate, I would be more inclined to place my money on the conclusions of the 2 000-odd scientists who contribute to its research than on the findings of a single political statistician with no background in climate science. I would also warn the public not to be distracted from the core issues by the bun fight caused by Lomborg and other sceptics.

I was offered a rare opportunity to board the *SA Agulhas* and travel down to the Prince Edward Islands in the sub-Antarctic where South Africa has a meteorological station. My mandate was to follow some of the work of a group of climate change researchers. There I discovered a place so entirely removed from the world we know here, with its littered pavements, congested highways and mono-crop agriculture. This remote mound of volcanic rock, poking out of the ocean and blasted year round by the ferocious westerly winds, is one of the few inaccessible wildernesses left on Earth. But even here the effects of pollution from half-way across the planet are changing the face of the islands. They are warming up and drying out and the plants and animals which live here – mostly stunted and slow-living after millennia of evolving in  such a cold and windy place – are changing too.

Part of my intention in this book was to tell the stories of those non-complicit life forms whose very existence is threatened by the adverse effects of centuries of fossil-fuel pollution. I did not intend to debate whether or not climate change is happening or whether it's our fault. There is a place to pit the dissidents against the mainstream scientists on the subject, but it would require a volume of its own. This book was also supposed to avoid the politics of climate change altogether because I wanted its pages to marvel at the wonders of nature rather than fuss over the complicated entanglements of the human condition. But I did not manage to stay out of politics altogether.

I have done what many scientists recommend: moved beyond arguing about whether or not climate change is happening and whose fault it is. There's no more time. Instead I have focused on what the outcome will be if we continue to turn up the thermostat. We have gone way beyond a point where we can afford more bickering, finger pointing and political machinations. It's time to find solutions. If we do not, it really will be 'the end of the world as we know it'. It could even bring about the extinction of the human species, along with countless other plants and animals which are unable to adapt fast enough to survive the environmental changes. Humanity will cease to exist – and it won't be the gentle falling asleep of an aged person who has run his or her course and peacefully expires while asleep. It will be the agonising, wretched death of a youthful person, snuffed out cruelly before his or her time. There will be no soft focus and roses at the close of this story. The Earth will recover. Life has shown an astonishing resilience in the past. But you and I, and our grandchildren, might not be part of that future if action is not taken *now* to cut back on greenhouse gas emissions.

Until such a time, life carries on. Each day you and I scurry to work and worry about making ends meet; governments draft bills and table motions around numerous agendas; climate change scientists feed numbers into computers and mull over the sometimes impenetrably complicated results. While all this goes on, plants and animals outside of this paradigm wake or retire with the Sun every morning and go about the business of surviving. Sometimes they must contend with sprawling cities, roads carving up their breeding grounds, agriculture unrolling in an endless sheet of mono-crops, and livestock grazing down their familiar turf. Then there is pollution, poison, aggressive invasive species from distant lands, and rivers dammed up and diverted – flushed with fertilizers, pesticides and soil from naked lands. Some species walk a tightrope of survival almost daily; others adapt so well to their changing world that they threaten to become a problem themselves. Now all these species must contend with a changing climate too.

I have taken some of the predictions put forward by South Africa's scientific community and put a face to them: elephants, an expansive savanna,  a spread of reef and croaking frogs, among others. Hopefully some gritty reality has emerged from the graphs and maps churned out by the complex algorithms. And a strong sense of responsibility.

# Chapter 1

Earth: a swirled island in the vacuum of space
Image courtesy of NASA

# The longest
# journey

*Earth's aura, our air, is unusual. It turns out on examination to be unlike any other atmosphere we know of. It has strange properties that make it necessary to include the air we breathe amongst Earth's growing collection of rare and unreasonable things.*

Lyall Watson, *Heaven's Breath,* 1984

Two figures were sitting on a bench under the wrinkled face of Table Mountain early one Sunday morning when the elder of them paused in conversation and pointed to a nearby tree.

'See that tree?' he asked.

I nodded, curious.

'Well, let that be a lesson to you . . .'

Every now and then I try to do just that, to look at things a little differently. Take the sky. If you look up after sunset when the stars are pulsing against the night like some kind of interstellar Morse code, you'll see what I mean. Hopefully there will be no clouds to obstruct your view and, if you are fortunate enough to be in the countryside, you won't have to contend with ambient city light to dilute the impact. But look at those stars and imagine for a second that you are a passenger in a car, forehead pressed against the window while you stare out at a desert landscape as you speed by.

You are, in fact, doing just that. Your vehicle for this particular journey is an extremely large rock that is hurtling through space at 106 000 km per second. Like a Formula One car on the racing circuit, you are stuck to one route and one route only. But it's a long one – each lap is a 938 900 000 km loop, which takes you a year to complete, so you shouldn't get bored with the repetition. Unfortunately the scenery out there won't change much. Your window, in this case, is a thin layer of gas between you and the chilly outdoors. It protects you from the kind of debris that usually splatters across your everyday space-enabled windscreen.

By now you should have slipped into the contemplative thousand-metre stare that grips most passengers on a long-haul journey. Let me draw your attention back to the book for a while to consider this: it's a marvel that you are reading the words on this page, and not just because you have been spared a bad case of travel sickness. There is a great deal more that should impress you. It's more than the fact that you were born with organs complex enough to pick up the light reflected from the page, or that you share with your species a highly evolved brain able to translate the text you see before you into a narrative. It's more than the knowledge that you are recycled from the constant turnover of billions of years of cosmic dust, which more than likely once resided in a *Tyrannosaurus rex* or an early-Cambrian trilobite.

That's radical. But what's also pretty amazing is that while you are thinking about how splendid you are – which by now you should be – you are sitting at the bottom of an ocean of air that your body is unconsciously taking in and expelling in a steady and constant rhythm. Your heart is chiming in, carrying the extracted oxygen to every cell in your body, including your brain, so that you can comprehend the words and steer your hand to turn the next page.

Until now you probably weren't thinking about the fact that what you are sitting in is mostly nitrogen. In fact, 78 per cent of each breath you draw is nitrogen, diluting the 21 per cent oxygen content that keeps you conscious and functioning, along with a smattering of water vapour, carbon dioxide, carbon monoxide, methane, helium, argon, neon and a few other gases[1]. If there were a fraction more oxygen in the gaseous cocktail, you would not be sitting so contentedly. If there were about four per cent more oxygen[2], or a little less than the fraction of methane[3], this page would spontaneously combust. The world would have been a fiery and volatile place. If you had been able to evolve at all, it might not have been into the soft-skinned animal that you are. These pages would not have happened either because the forests from which they are made might have been torched before you could say 'Gutenberg press'.

There's a symmetry in the air around you that is beautiful, poetic, invisible and entirely underrated. If we did not have the fraction of carbon dioxide (0.037 per cent of the total) wrapped around our planet like a shawl, life as we know it would not have stood a chance in a frozen wasteland.

There's a minute amount of ammonia too. Every time lightning slices through the sky, it causes a chemical reaction in the atmosphere as nitrogen and oxygen combine to form nitrous acid. Without the alkaline effect of ammonia to neutralise the acid, the biosphere would slowly scorch itself to death. The ridiculously thin layer of ozone at 20 to 25 km[4] up keeps us all from being baked in our skins and our cells from being mutated by DNA-scrambling ultraviolet light.

After 4 600 million years, an atmosphere has evolved in such glorious and unlikely equilibrium that it makes me want to breathe a little deeper on the deliciously tasteless stuff.

Throughout your lifetime you will suck in around 250 million litres[5] of air as you go about the business of living. In that time, you may occasionally want to ponder on this: just how unlikely is it that the 'third planet from a minor star'[6] enjoys a series of physical characteristics that achieved a breathable atmosphere, one in which billions of years of evolution produced a species that adapted the world around it so masterfully that it could escape that very atmosphere in a chunk of flying metal controlled by messages conveyed in a binary code of 0s and 1s?

• • • • • • •

At first there was nothing. Nothing and nowhere. There was neither light nor dark; there was neither inside nor outside. There was neither form nor even a void; no time, no space. Nothing at all. Except for the singularity: a mathematical point with no volume from which exploded all the matter and energy ever needed to sculpt our universe. In a brief moment – faster than the flick of a hummingbird's wing – the singularity erupted in a cataclysmic explosion in which the rudiments of everything that make up you and I were spun out across an inflating universe. All matter and all energy were hurled outwards, a massive exploding canvas upon which life could eventually begin its work. Time and space came into being. Within a billionth of a second it had spread itself out across hundreds of millions of kilometres[7] and suddenly there was past, present and future[8]. This was the birthday of our universe, about 13 000 to 14 000 million years ago.

As the universe expanded and cooled over the next 300 000 years, the fragments gathered together to form the first two items on the periodic table: hydrogen and helium. A fraction of lithium and beryllium found their way into existence too. It took several hundred million years for the primordial gas clouds of hydrogen and helium to cool. Gravity inherent in these clouds began to tug massive balls of gas together, which eventually formed the first stars, globular clusters and galaxies. These early stars were the laboratories in which the heavier elements – up to iron on the periodic table – were cooked, at temperatures of 15 million degrees and more.

As these early giant stars became supernovae – cataclysmic explosions, which destroyed the massive stars, leaving behind small dense neutron stars or black holes – the elements beyond iron such as gold, tungsten and uranium formed in these cosmic super-labs. The supernovae remnants became clouds of gas and dust that, in turn, formed new stars orbited by cooler clusters of matter. These later spun together to become planets. This cyclic process continues to this day – stars dying to form new ones[9].

The wonder of our existence is that of all the suns out there, with all their attendant planets, moons and circling debris, at least one turned out to have the right criteria for life as we know it to emerge. There may be more such planets out there – until we have looked we cannot be sure. But in the words of Arthur C Clarke: whether or not there is life out there, the idea in either case is quite staggering. Of course, we know now that our star is a minor one in a galaxy filled with a few hundred billion stars. The fact that these are several generations down the family tree, after aeons of solar births and deaths, should make us realise that eventually at least one planet had to stumble upon conditions favourable for some kind of life.

There are various additional things to be thankful for about our solar system as you contemplate life over your morning cup of coffee:

**The Sun:** Without the Sun, life does not even leave the starting blocks. We need its gravity to keep us safely spinning through its space, which technically comprises the outer reaches of a nuclear reaction. Then there is the energy it sends us. The Sun's surface temperature is 5 800°C, which radiates out into the solar system. The small amount that reaches us 150 million km away[10], about 8.3 minutes[11] after departure from the Sun, warms the planet and drives the global climate. It provides energy for plants, upon which all animals are dependent for food, oxygen and a place to live. The Sun accounts for 99 per cent[12] of our solar system's mass, so it's due a great deal of respect. It's going to burn itself out in 5 000 million years. Thankfully you and I won't be around to witness this because things on Earth will have become a little uncomfortable by then.

**Distance from the Sun:** All the inner planets in our solar system started out with a rudimentary wrapping of mostly hydrogen and helium, but those planets too close to the Sun had no chance of keeping theirs. The secondary atmospheres that formed around the smaller planets from gases released through the planets' crusts also ran the risk of being boiled off by the bombardment of solar radiation. Mercury's daytime maximum reaches a towering 430°C[13].

But, like most things, it's all relative; in this case, to the . . .

**Size of the planet:** A planet's size and mass determines its gravitational hold on the gases in its atmosphere. These gases are made up of atoms and molecules that jostle about at speeds which are dependent on their temperature: the hotter the gas, the faster the molecules move. Once this speed is great enough, the molecules will escape the gravity of a planet, as if 'boiling' away. This speed is known as the escape velocity. Lighter gases such as hydrogen and helium are the first gases to reach escape velocity and to 'leak' away from the planet. Mercury, for example, has a gravitational pull only 40 per cent that of Earth's[14]. This means that molecules in the atmosphere, exposed to such ferocious solar heat, reached escape velocity swiftly and bade their little planet farewell. Almost the size of Earth, Venus managed to hold onto its atmosphere. Venus's atmosphere consists mostly of carbon dioxide, is 90 times heavier than ours, and keeps the planet at a constant 480°C. Mars – just one stop out from us – has a gravity only slightly greater than that of Mercury. Therefore Mars, too, has lost most of its gaseous shroud, leaving behind only the heavier gas – carbon dioxide. The temperature on Mars vacillates between minus 120°C and 25°C[15] but averages out at minus 42°C[16]. Not quite a holiday in the Cape.

Earth found itself with a perfect balance among distance from the Sun and the right size, mass and hence gravity to keep its atmosphere intact. It produces temperature extremes of minus 70°C to a sweltering 55°C but with some eminently liveable moderates in between. This may not have been the case had an object the size of Mars not smashed into Earth soon after its conception, about 4 500 million years ago. The massive collision blasted rock from the Earth's crust out into orbit where gravity kept it circling and moulded it into our silent companion, the Moon. If significantly less material had broken away, the body it formed might have dropped back down to Earth. If it had been bigger, it would have flung itself out of Earth's orbit. Either way, we would not have the lunar tidal effects we experience today. Earth would also have been bigger with a stronger gravity. If we had been fortunate enough to evolve, we might have turned out to be short, squat and built for higher pressure.

**Atmospheric composition:** Of course it's no good having an atmosphere if it's not something that future life can breathe. After the first hydrogen-helium concoction, which was 'blasted away by impacts and solar wind'[17], Earth created for itself something rather more sustainable. Early volcanic activity filled the void around Earth with nitrogen, oxygen and water vapour, which condensed to form the oceans, from which ultraviolet light split oxygen to make the ozone layer.

**Greenhouse effect:** The Sun's radiation penetrates the atmosphere in short ultraviolet waves. Some bounces back and some is absorbed by the Earth to be re-emitted in longer infrared waves. Small amounts of so-called greenhouse gases – carbon dioxide ($CO_2$), methane ($NH_4$), nitrous oxide ($N_2O$) and various others – absorb this infrared heat and hold on to it long enough to keep the planet warm. Without greenhouse gases to 'catch and hold energy, banking it against lean times'[18], Earth would be 30°C to 40°C colder than it is now[19].

**Life:** The system needed something else though – life. Without the evolution of organisms, our atmosphere might have ended up something like that of Venus[20]. About 2 700 million years ago, the earliest photosynthesising organisms began producing oxygen in bulk. The formation of the ozone layer meant protection from harmful ultraviolet light, which makes a mush of DNA.

**Convection:** Without heat transfer, life as we know it would be squeezed into a narrow belt of comfort at about '38° north and south of a boiling equator'[21] while the rest of the planet would be a frigid hell. Heat from the Sun warms the surface of the planet: air at ground level heats, expands and rises. Cool air rushes in to replace it, creating massive bodies of perpetually circulating air. Because most heating takes place at the equator, the air rising here moves polewards before it cools and sinks. As the air moves up and away from the equator, the Earth's spin deflects it to the left in the Southern Hemisphere and to the right in the Northern Hemisphere[22]. This creates 'cells' whose actions, together with the Earth's spin, then drive the ocean currents. These two processes are so interrelated that they are often regarded as one system. They constitute the Earth's thermostat and keep things at a liveable average of 15°C.

There you have it – the start of a self-sustaining life support system into which you and I were born billions of years later. Spinning through the vacuum of space, along with so many other apparently lifeless forms, this planet really is a marvel with its thin transparent layer of gas that can:

> Edit the Sun [and] provide shelter from the showers of meteorites that hammer on the outer edges of the atmosphere, like the random noise of rain on the roof at night . . . [can] feel the fluctuations in pressure that set particles in motion, spreading agitation in trains of waves, carrying news and information in the form of sounds.[23]

But there is nothing soft focus and mushy about it all. The same air – all 5 600 million million tonnes[24] of it – that cushions the blow of raindrops[25], can deliver enough force to level buildings once it gathers enough momentum. Nature is no pussycat – just look at the brutality of the food chain – and it has gone through its own share of mood swings, not least the global climate. We know that climate has not been constant – the fact that Antarctica has fossils of coniferous forests dating back about 100 million years when the continent was in its current position over the poles[26] proves this. Glacial activity over parts of South Africa tells of another extreme. And, at least three times in its life, our spinning orb has been almost entirely frozen over – a condition referred to as Snowball Earth.

Global climate is driven by Earth's thermal state, which is dependent on a simple principle of physics – the balance between heat in and heat out.

## Heat in

The year 1816 was known as the year without summer. Well documented in several sources, the winter chill lingered, seedlings froze, crops failed, livestock died and people starved. In the north-eastern United States, 1816 became known as 'eighteen hundred and froze to death'[27]. Only later was a link made between an average temperature decrease of a miniscule 1°C and a volcanic eruption halfway across the globe. The eruption of Tambora on the Indonesian island of Sumbawa put enough dust and ash into the atmosphere to prevent a fraction of the Sun's heat making it through to Earth. People starved as a result.

Like moths around a flame, most of our heat comes from the Sun. Of that energy reaching our outer atmosphere, just over half makes it through to the surface of the Earth. The angle at which the heat strikes the atmosphere influences how much makes it through. An acute angle, closer to the equator, means less atmosphere to pass through and more heat penetration; an obtuse angle closer to the poles means more atmosphere to interrupt the heat so less reaches the ground. For this reason the poles are colder than the equator and the same principle shifts the mercury between winters and summers in the two hemispheres. The remaining energy is either reflected back out to space by clouds, dust or snow on the Earth's surface or is absorbed by clouds. Some heat is trapped near the surface by greenhouse gases, keeping the planet snug and warm, except for those few extreme places.

'Heat in' changes as the Sun pulses. This is associated with the tidal action of the Sun and occurs in 22-year cycles: in fact, two 11-year cycles during which solar activity reaches a maximum and a minimum[28]. Such activity has been written in the growth rings of Ponderosa pines in Arizona, which show growth spurts during warmer periods. An increase in solar activity is associated with increases in temperature and more rain in the Northern Hemisphere; less activity produces a decrease in temperature and a decline in rainfall there.

The Earth's path around the Sun is not entirely as rigid as suggested earlier. Actually it has a disconcerting tendency to wobble and stretch its orbit, which has a significant impact on the amount of heat that enters our system. The Earth's orbit follows three distinct cycles. These are grouped together as the Milankovitch cycles, named after the Serbian scientist who calculated their oscillations many decades before the scientific community finally accepted them.

Milutin Milankovitch made a correlation between the clockwork regularity of ice ages and the Earth's orbiting cycles. He first found that Earth's path around the

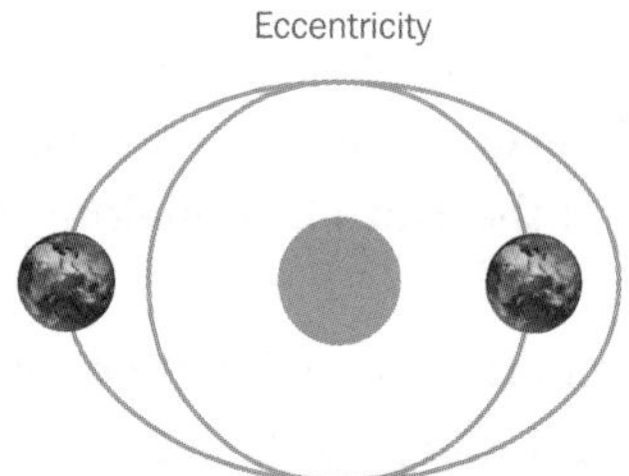


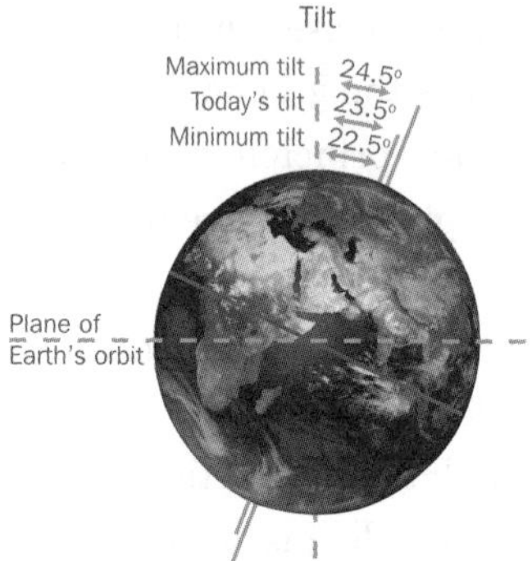


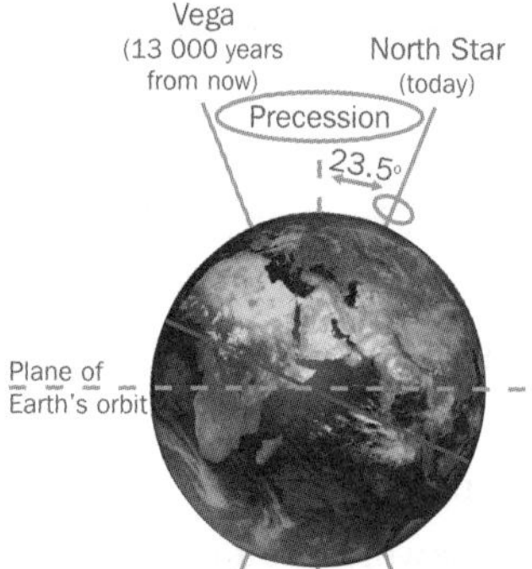


Sun is not constant, but stretches its elastic orbit between a circular and elliptical course every 100 000 years. This changes our distance from the heat source and has a profound impact on global climate, triggering glacial and interglacial cycles that correspond to the 100 000 year cycle.

It did not end there, Milankovitch discovered. Another two cycles repeat more regularly: every 42 000 years the Earth's axis tilts between 21.6° and 24.5° on its equatorial plane, while every 22 000 years the axis wobbles like 'a top about to fall'[29]. The latter movement, known as the precession, wobbles the planet between two phases: every 22 000 years, when the angle of precession puts the Northern Hemisphere closest to the Sun during its summer, it will receive the most concentrated sunshine[30], mirrored by warmer temperatures for the region; 11 000 years later, when the precession is at the opposite side of the cycle, the Northern Hemisphere will be further from the Sun during its summer time and thus will receive less light. Temperatures in that region will drop. A similar pattern will be seen in the Southern Hemisphere, but in reverse.

Over the past three million years, these regular changes in the amount of sunlight reaching the planet's surface have produced

The Milankovitch cycles: the Earth's path around the Sun stretches, tilts and wobbles.

a long sequence of ice ages (when greater areas of Northern Hemisphere continents were covered with ice) separated by short, warm interglacial periods.[31]

## Heat out

How much heat the Earth retains depends on two things – the amount of greenhouse gases present in the atmosphere, and how light or dark the surface of the Earth is. It makes sense that if you increase the amount of greenhouse gases trapping heat, you will increase the amount of heat that stays in the system. It's like pulling on a thicker sweater – you trap more body heat and warm up.

Meanwhile, light surfaces such as snow and ice send a lot of energy back out to space, like tinfoil reflecting heat. Darker surfaces such as forests absorb heat. This so-called albedo effect is crucial in regulating the thermal state of things. The more ice there is, the less heat stays. Ice ages are believed to be triggered not by winter getting colder, but by summer not warming. If summer does not warm enough to melt the most recent deposit of snow, the white covering remains and keeps redirecting heat out to space, causing the planet to spiral into freezer conditions. But, for this to happen, the system has to be close to a tipping point.

Various other things shaped climate at a local level. The tectonic movement that pushed up the Himalayan mountains by smashing together the Indian continental and South-East Asian plates to produce an almost impassable barrier of rock and the highest summit on the planet, single-handedly created the South-East Asian monsoon. This movement, and the formation of the Isthmus of Panama, are believed to have disrupted ocean and air circulations sufficiently to trigger ice ages after their formation[32].

The system ticks over like clockwork, producing oscillations between warmer and cooler periods. Life on Earth evolved to this steady and reliable cadence.

# Chapter 2

Even remote species such as the wandering albatross will feel the heat
Photograph © Leonie S Joubert

# Tuning out the life support

*This is not a belief system, it's an observable scientific fact.*
Scientist Nicholas Bates, on elevated $CO_2$ concentration in
the ocean's waters – *National Geographic*, September 2004

Every now and then something cataclysmic happens. Take the dinosaurs. They departed in a hailstorm of fire and molten rock. The only evidence of their sudden exit is a layer of soot-laden clay no thicker than the width of your thumb. Buried deep in the bowels of rock are the scattered ashes of their incinerated remains, a 65 million-year-old crypt wrapped around the planet.

The age of this bird- and lizard-hipped reptilian family came to an end when a 10 km-wide meteorite struck Earth, releasing the energy equivalent of 100 trillion tons of TNT[1]. The impact vaporised the meteorite, liquefied rock and blasted a massive disintegrated chunk of the Earth's crust into the atmosphere. Shot back into the heavens, the high-velocity rubble spread out into a cloud up to 200 km wide. Beneath it, the Earth continued spinning, wrapping a lethal ribbon of debris around itself. Gravity began pulling the wreckage back to the surface at speeds of 7 000 to 40 000 km per hour[2]. The heat from this massive re-entry rendered many animals unconscious, a blessed relief from the approaching fury. In that instant, even marshy vegetation dried to a crisp before bursting into flame. Massive swaths of the planet were engulfed in wildfires. Tsunamis and earthquakes added to the turmoil. Dust and soot snuffed out any sunlight, creating a nuclear-type winter that lasted for months. Photosynthesising plants wilted and died. The food chain collapsed. Two-thirds of all animal species died, including the dinosaurs which had dominated the planet for 100 million years.

It took hundreds of thousands of years for the rudiments of a functioning ecosystem to start ticking over again and several million more for the diversity of life to get back to pre-impact levels. All that remains of the Chicxulub impact is a 240 km-wide crater hidden beneath Mexico's Yucatan Peninsula[3], the tell-tale layer of sooty clay, and the childish fables of the dragons of yore. Messing with life at a planetary level has been the preserve of big league events such as meteor strikes, volcanoes and earthquakes. The human species, astoundingly, has elevated itself to rank alongside these, achieving the 'dubious distinction' of becoming a 'geophysical force on the planet'[4].

The composition of the atmosphere has changed constantly over time, as has the thickness of the sweater of greenhouse gases, swinging between different states of equilibrium. Carbon is recycled constantly. From the atmosphere, where carbon exists as carbon dioxide, it is absorbed by plants which breathe some back out but convert the rest to carbohydrates, fats and proteins. Herbivores eat the plants and store these elements in their tissues or expel some through breathing. Both release carbon back into the atmosphere when they die and rot. This is how the bulk of carbon circulates. Occasionally carbon is banked away in ancient forms such as coal, gas, crude oil or limestone. Forests and plankton-rich oceans are regarded as 'carbon sinks' or places for the short-term storage of carbon. They are crucial in keeping balance in the system. Methane is produced mostly through the rotting of soggy vegetation, such as the kind you would find in wetlands, marshes, bogs and the like. Herbivores also produce their share of the gas by digesting their greens … and I'm not talking about belching. Nitrous oxide is emitted by bacteria present in the soil and by the oceans.

Humanity has interfered with these cycles. We've dug out ancient fossil fuels

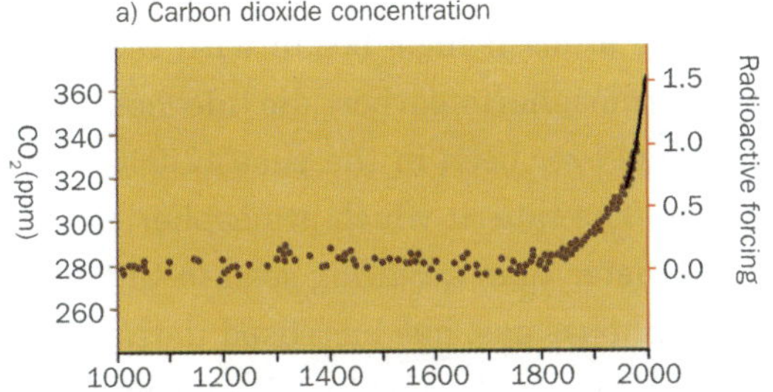


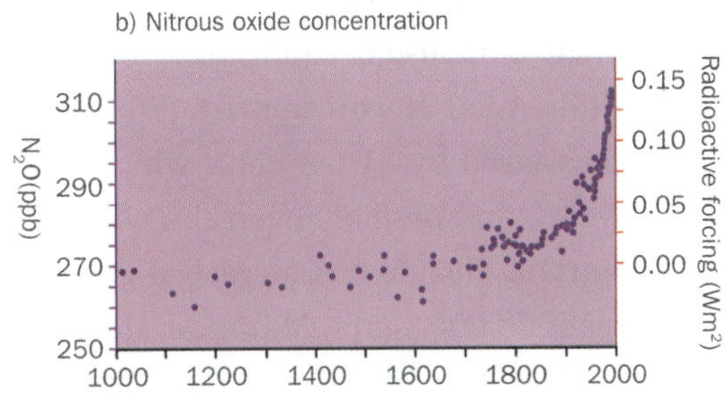


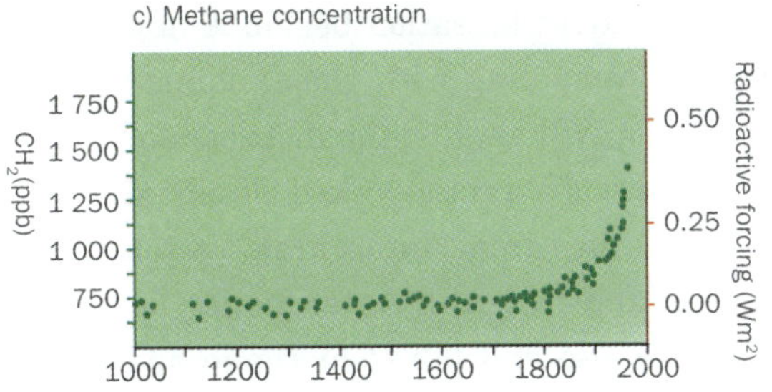


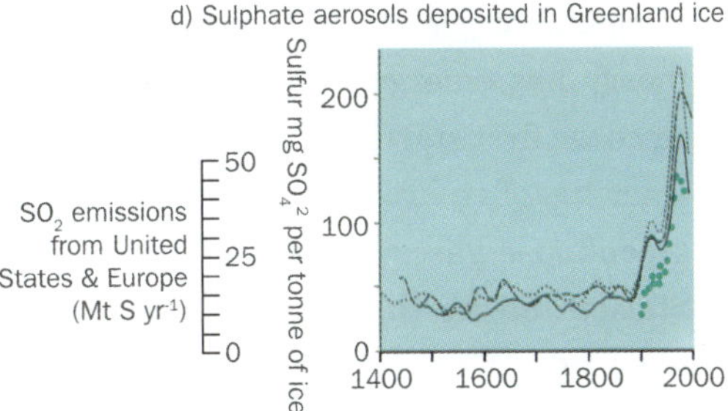


Graphs illustrating the increase in concentration of atmospheric greenhouse gases a) carbon dioxide, b) nitrous oxide, c) methane and d) sulphate aerosols since the beginning of the industrial era.
Courtesy United Nations Intergovernmental Panel on Climate Change

and burned them. We've cut down forests and burned those too. We've inundated millions of hectares with rice paddies, creating artificial wetlands that churn out methane. And the global cattle herd has grown larger and larger as we've farmed meat for the masses. Every one of us, no matter how we live or what we eat, is responsible at some level for contributing to this change. By looting carbon banks and shunting their contents back up into the atmosphere, we have altered the heat-in:heat-out equation. Since the mass burning of fossil fuels began with the Industrial Revolution in 1750, the amount of atmospheric carbon dioxide has increased by 31 per cent, which is as high as it has ever been in the past 42 000 years and possibly even the past 20 million years. The rate at which atmospheric carbon dioxide has increased is also significantly faster than anything seen in the past 20 000 years. Methane, the more potent greenhouse gas, has increased in the atmosphere by 151 per cent during this time and nitrous oxide is up by 17 per cent. Clearly this is going to enhance the greenhouse effect[5].

Furthermore, there may be reason to believe that our interference with the atmosphere goes back thousands of years before industrialisation began. A new theory has emerged which implicates humans in tampering with global climate since we first started tilling the soil. Marine geologist William Ruddiman, professor emeritus of environmental sciences at the University of Virginia, looked closely at greenhouse gases extracted from a 3km ice core drilled from Antarctica's Vostok Station during the 1990s. Air bubbles, captured and frozen in Antarctic ice for hundreds of thousands of years, confirmed that a rise and fall of methane and carbon dioxide levels coincided with the complex Milankovitch orbital cycles. Atmospheric methane, Ruddiman argues, is naturally higher during warming interglacial periods. A warmer climate means more abundant plant growth in wetlands, translating into more material to decompose and emit methane. The reverse is true for cooler periods near the onset or exit of glacial periods where wetlands produce less plant material and hence less methane. The flux precisely matches the 22 000 year short-term precession in the Milankovitch cycle.

Carbon dioxide concentrations in the atmosphere also climb and fall in response

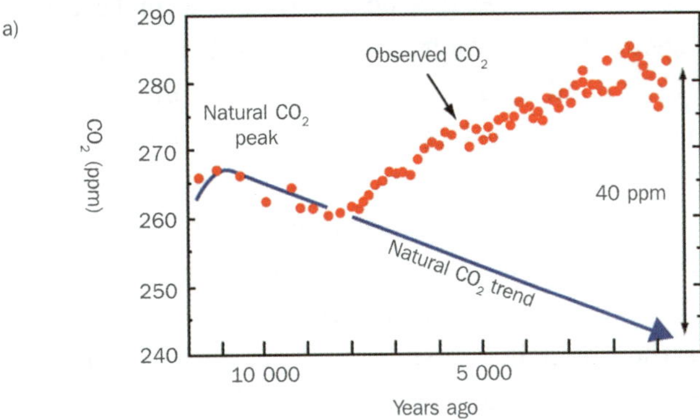


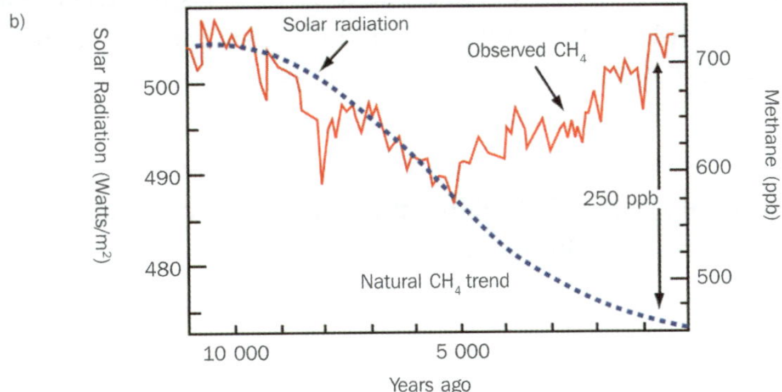


Atmospheric  a) carbon dioxide and b) methane should have been on the decrease for the past 10 000 years, but both show sudden increases at 8 000 and 5 000 years ago respectively.
Courtesy Professor William Ruddiman

to a more complicated series of triggers in the Milankovitch orbits. By 10 500 years ago, methane and carbon dioxide levels were at a natural peak, typical of previous interglacial periods on record in ice cores. In keeping with the normal cycle, both gases began to decrease and should have continued to do so until the present day as our system should currently be heading into a naturally cooler phase. But strangely, 8 000 years ago carbon dioxide reversed its trend and began to increase slowly. 'By the start of the industrial era, the concentration had risen to 285 parts per million (ppm) – roughly 40 ppm higher than expected from the earlier behaviour[6].' Similarly, methane reversed its own trend at 5 000 years ago in the ice core records and, by pre-industrial times, was 250 parts per billion (ppb) higher than it should have been.

Ruddiman attributes the anomalies to spin offs from agricultural practices by the newly settling *Homo sapiens*. About 8 000 years ago the farming of cereal crops had spread from the Middle East and into Europe. Evidence of silt washed into rivers testifies to land being denuded of trees to make way for crops. The loss of this important carbon sink, and the burning of those trees, could be a likely explanation for the altered course of atmospheric carbon dioxide. The initiation of rice irrigation in South-East Asia could explain the sudden corresponding increase in methane levels. 'Farmers began flooding lowlands near rivers to grow wet-adapted strains of rice around 5 000 years ago in the south of China . . . Historical records also indicate a steady expansion in rice irrigation throughout the interval when methane values were rising[7].' If Ruddiman is correct, then humans have been altering global climate radically for thousands of years, even with significantly more primitive means.

According to the Milankovitch orbits, Earth should be descending into an ice age in the next few hundred or, at most, the next few thousand years but enhanced global warming owing to agricultural practices may have made parts of the planet more favourable for our survival[8].

During the past 2 000 years there have been a few occasions where the $CO_2$ concentrations have dropped quite drastically, taking the global temperature down with them. Traditionally these occasions have been attributed to changes in output from the Sun or volcanic eruptions. Ruddiman thinks otherwise. He argues that deforestation linked to human agriculture had already been increasing $CO_2$ levels since 8 000 years ago. What would happen, his theory poses, if the human population took a sudden and massive dip? From 540 to 542 CE (Common Era) and again from 1347 to 1352 CE, a Roman-era outbreak of bubonic plague decimated between 25 and 40 per cent of the European population. The introduction of smallpox and other diseases to the Americas by European explorers after 1492 CE killed over 90 per cent of the pre-Columbian population, some 50 million people.

In many cases, population crashes of this kind were associated with people abandoning villages and urbanising. Agricultural fields then reverted to their wild state, with forests taking about fifty years to re-establish themselves in abandoned farmlands. These epidemics and their resulting decrease in population mirror the fluctuations in atmospheric $CO_2$ in the past 2 000 years. Computer modelling suggests that this recovery of forest was enough to allow sufficient absorption of $CO_2$ to push the trend towards cooling events. Once human populations bounced back from their crises, they reversed the trend to one of increasing atmospheric $CO_2$ as they began burning and slashing into forests again[9].

If this is the case, then we have been manipulating global climate to our benefit – we've kept things cosy and warm when we should be permanently wrapped up in our winter woollies. Dissidents in the debate use this point to argue that global warming is good because much of the cold Northern Hemisphere will become more habitable. As conditions warm in the north, greater areas will become suitable for agriculture, thus benefiting food security. The problem with this argument is that while the north is warming and possibly thriving, many parts of the south are collapsing under drought. But more of that later. Another flaw in this debate is that we are conducting a dangerous and uncontrolled experiment. We do not really know what will happen as we continue to add energy to the planetary system.

Whatever the reason for the changes in atmospheric composition, there are some significant consequences[10]. The average temperature has increased since 1861 when reliable weather records begin. The twentieth century saw an increase of about

0.6°C, probably the largest increase in the Northern Hemisphere of any century during the past 1 000 years. The 1990s was the warmest decade and 1998 the warmest year on record (although 2005 recently achieved a new record high). Daily minimum temperatures in the north have increased by 0.2°C between 1950 and 1993, about twice the increase of daily maximum temperatures.

This is how it is playing out in nature – ice is melting faster than snow can replenish it[11]. The Glacier National Park in Montana in the United States had 150 glaciers. Now there are fewer than 30 and these have shrunk by two-thirds[12]. Kilimanjaro's white top has decreased by 80 per cent in less than a century. The central and eastern Himalayan glaciers may disappear by 2035. Artic sea ice has thinned by 40 per cent during the late summer and early autumn months[13] and the loss of sea ice is posing a dire threat to the survival of polar bears, seals and Inuit communities[14]. The Arctic is warming faster than anywhere else because of the albedo effect – more ice melts, therefore greater amounts of dark ground are exposed to absorb rather than reflect the Sun's energy, and the regional temperature climbs.

In Fairbanks, Alaska, permafrost is melting. Roads built level on solid, frozen ground now look like crumpled carpets, becoming higgledy-piggledy as the ground beneath them turns soggy. Trees that were once supported upright by the frozen ground have started lolling drunkenly[15]. Further south, on the underbelly of the planet, the Larson B ice shelf – a massive slab of ice floating off the eastern edge of the Antarctic Peninsula – disintegrated in 2002 as 12 500 square kilometres of ice broke off into icebergs and floated away into the Weddell Sea. This happened in as little as 35 days after 10 000 years of stability[16]. Steady warming by 2.5°C has been recorded here over the past 50 years[17].

In August 2003 Britain's Department of Health issued guidelines on how to cope with an impending heat wave. When it struck, the mercury hovered in the mid-30s before climbing beyond anything ever recorded in Britain. Poultry farmers later began switching to maize and sunflower crops because of the mortality rate of their birds in the heat. Rail services ground to a halt when temperatures threatened to buckle the tracks. Wine farmers from the Champagne region of France began

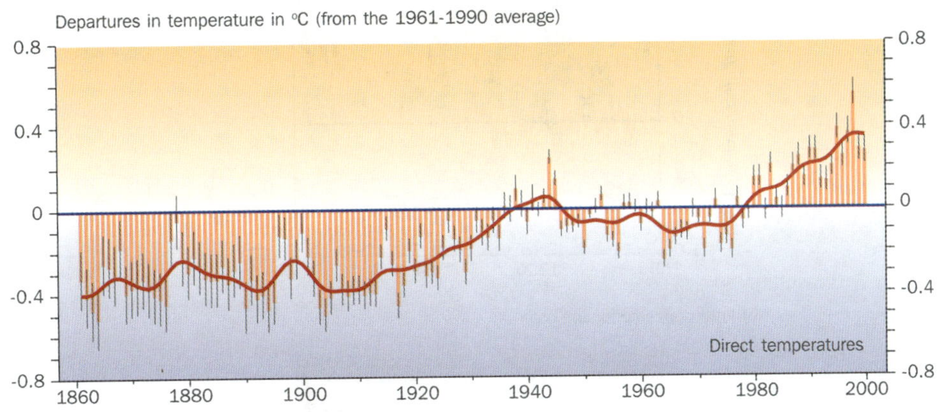



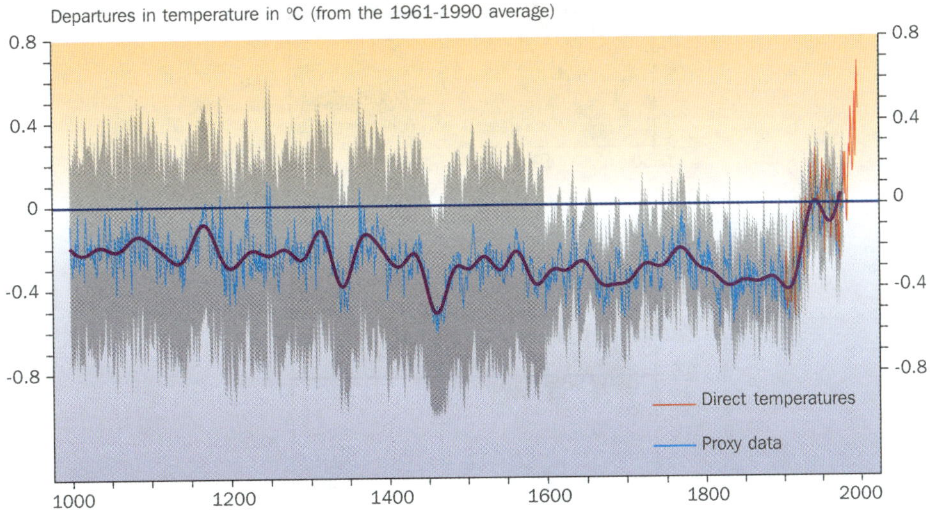



Variations of the Earth's surface temperature for the past 140 years (global) and the past 1 000 years (Northern Hemisphere).
Courtesy United Nations Intergovernmental Panel on Climate Change

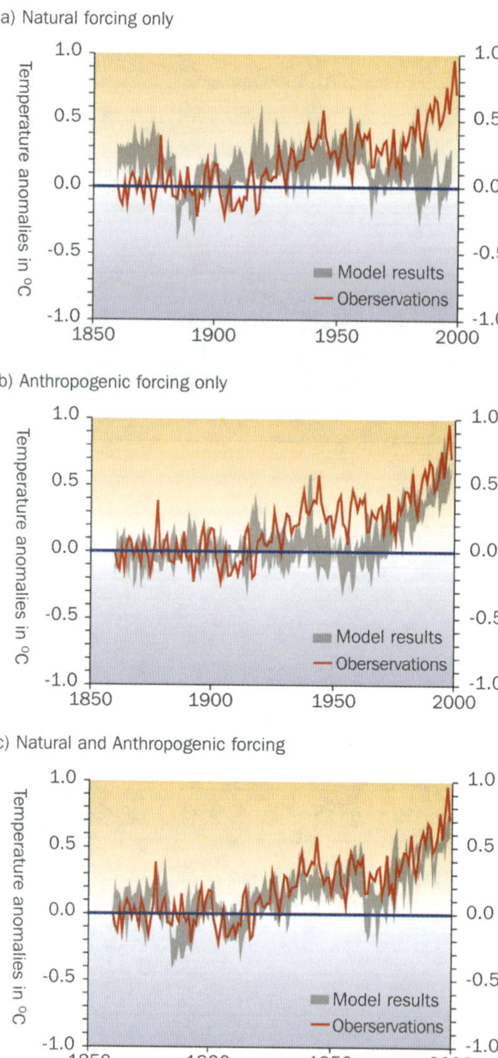


Comparison between modelled and observed temperature rise since the year 1860. This is a means of testing the accuracy of computer models in predicting future climate changes and atmospheric ghg change.
Courtesy United Nations Intergovernmental Panel on Climate Change

scouting for property along the chalky slopes of south-east England as steady warming made the area look more and more agreeable for producing the crisp, acidic wines needed for a good bubbly. Across the channel, the rest of Europe was struggling with the same record-breaking heat wave which pushed up temperatures by 3.5°C higher than average, causing what was probably the hottest summer that region had experienced in 500 years. During a two-week period, 22 000 to 45 000 deaths in the region (14 000 of which occurred in France alone[18]) were caused by this weather phenomenon, mostly among the elderly and infirm where heart and respiratory complications owing to heat stress eventually proved fatal[19].

Heat wave episodes will continue into the future and will probably be worse in cities where urban heat islands – in which heat is trapped over cities due to a complicated interaction of evaporation processes, impervious surfaces and changed vegetative cover – could push up the temperature by 5 to 11°C compared to that experienced outside of the city. Dr Myles Allen[20], at Oxford University's physics department, wrote in the journal *Nature* that the heat wave which swept through Europe was the result of an anticyclone where greenhouse gas emissions had 'loaded the weather dice . . . increas[ing] the risk of an anticyclone causing a heat wave like that of 2003 by around a factor of four'.

The World Health Organization and the United Nations Intergovernmental Panel on Climate Change predict that climbing temperatures and changing rainfall regimes will increase the spread of infectious diseases such as malaria and dengue fever. The incidence of diarrhoeal disease and salmonella-related food poisoning can also be expected to climb, while changing weather patterns will undermine food security, leading to malnutrition and increased vulnerability[21]. The World Health Organization estimates that, since the 1970s, over 150 000 deaths every year can be directly linked to climate change[22].

Then there is the rising of the tides. The inhabitants of the tiny Pacific atolls of Tuvalu are accustomed to having sea water well up around their feet. Literally. The occasional normal upwelling of tides – where sea water rises up through the sponge of coral rock upon which the people here live – gathers in puddles on the ground before becoming rivulets as water inundates gardens, airstrips, roads. This upwelling is happening more and more frequently[23]. Because most of the nation is less than 50 cm above sea level, there isn't much margin for error for the people of Tuvalu – they are facing a national crisis. The Maldives and the islands and atolls of Kiribati face similar inundation.

In March 2004 a hurricane ripped through Santa Catarina province in Brazil. At 145 km per hour it was an entry-level hurricane – nothing like the category five (a meteorological euphemism for 'devastating') Hurricane Ivan which tore through the Gulf of Mexico later that year at 270 km per hour. What made the relatively mild-mannered Hurricane Catarina special, though, is that it wasn't supposed to exist at all. It emerged out of the cooler South Atlantic Ocean. Hurricanes form in the 'doldrums' of the ocean between latitudes 10 degrees north and 10 degrees south where prevailing low pressure troughs notoriously used to leave sailing ships of yore limp-sailed and stranded at sea. Hurricanes need a sea surface temperature of 27°C or more in order to form. Hot, moisture-laden air is forced upwards, followed by colder air rushing in from the periphery, creating a storm with high winds circling inwards towards a central 'eye'. In the Atlantic, these conditions usually only occur north of the equator. The Hadley Centre for Climate Prediction and Research, part of the United Kingdom's Meteorological Office, said Catarina was the first ever recorded hurricane in the traditionally cooler South Atlantic.

It has long been understood that a rising sea surface temperature will drive bigger and more ferocious hurricanes. Evidence suggests such storms have been growing longer lifespans and greater intensity for three decades already – linked directly to warming seas in the area – and the trend is expected to continue into the future[24]. And if the horror that Hurricane Katrina wrought on the Mississippi Delta in 2005 is anything to go by, a community caught weak or ill prepared can be crippled by a natural disaster of this magnitude. The ongoing changes in climate are global and they appear to be gaining momentum.

'Weather' refers to the daily conditions in the atmosphere around any one place – temperature, wind, cloudiness, rain showers, humidity. These are usually measured in surface variables such as temperature, wind and precipitation. How this plays out over decades – the World Meteorological Organization defines this period as 30 years – produces an average weather that can roughly be described as the general 'climate' of a region. Climate change means a shift in the overall system, which can happen over many decades or millennia, and is reflected in a shift in the average. Long-term climate change ultimately plays out in changes in our daily weather. Since different parts of the planet enjoy different kinds of weather, how climate change manifests itself will be different in each region.

In 2005 one side of the planet was laid waste by the hurricane that tore into the Mississippi Delta. On the other side of the planet, east African countries buckled with hunger as droughts left rural grain stores empty. These kinds of weather anomalies have happened throughout history so there's no reason to prophesy that these two events are harbingers of worse to come. But, as the scientific community explains, ongoing weather anomalies around the planet (the extremes, not necessarily the norms) piece together like a jigsaw puzzle with an emerging image that supports computer-generated climate modelling.

I know it's difficult to swallow long-range predictions by climatologists. After all, how can we believe them when the local meteorological office cannot guarantee the short-term forecast for Saturday's cricket? Yes, it's difficult to predict exactly whether tomorrow's maximum will be 35°C or 33.5°C. But the weather people are usually right about the general trend. And yes, sometimes they even get that wrong. Just because they are a little off occasionally does not mean they are peddling snake oil. Predicting tomorrow's weather is not the same as guessing the next decade's climate. Climate change modelling is extremely complex – even the most confident scientists admit that. But as computer modelling becomes more sophisticated, and the binary codes behind them ever more potent, scientists can predict with greater confidence the trends we are likely to see in the next one hundred years.

What will the world of our future look like? Hell won't freeze over, but Europe certainly might. It sounds counter-intuitive – an ice age in a warming world – but it could happen. The United Kingdom and Western Europe have a fairly mild climate compared with other regions at a similar northern latitude. Thanks is due to the Gulf Stream, a warm body of water that delivers about one million billion watts of heat[25] from the Gulf of Mexico across the Atlantic and past Scotland. Without any of the heat delivered by the Gulf Stream, Britain would be 5°C colder than it is. The Gulf Stream is a massive body of warm, salty water that cools and sinks when it reaches the North Atlantic, before starting the long haul back to the Antarctic along the ocean floor. It's part of the larger thermohaline circulation that distributes energy around the globe and drives the planet's climate.

A slumbering crisis lies at the place where this body of water cools and sinks – Greenland. It's the second largest icecap in the world and it's melting fast. Oceanographers fear that the amount of fresh, cold water flushing off Greenland into the head of the Gulf Stream could halt this conveyor belt – it has happened in the geological past. Significantly, a decrease in salinity has been recorded since the 1960s. A recent measurement was taken of the circulation process of the Gulf Stream across the 25° N line of latitude in the Atlantic Ocean and compared with similar measurements from 1957, 1981, 1992 and 1998. The journal *Nature* reported that this showed a slowing of the circulation of water by almost a third between 1957 and 2004[26]. Scientists now predict a 50 per cent chance of the Gulf Stream being halted. While it might not happen within the next hundred years, it could change the face of continental Europe and the British Isles when it does. You may want to stock up on your favourite Bordeaux now before the frost sets in at some of wine's most famed vineyards.

Melting ice and the physical expansion of a warmer ocean pushed sea levels up by 0.1 m to 0.2 m during the twentieth century. Anything from 0.11 m to 0.77 m is expected in the next century[27]. More than a hundred million people live within a metre of the mean sea level. Consider what this means in terms of food security for communities that grow cereals or vegetables just shy of the coast when salt water begins leaching into the water table. Consider the fate of the Nile Delta and Bangladesh. Louisiana's coastline is moving inland at close to a metre per century. Britain's National Trust, which owns nearly 900 km of coastline, has resigned itself to losing massive chunks of national heritage to the advancing tide. Meanwhile,

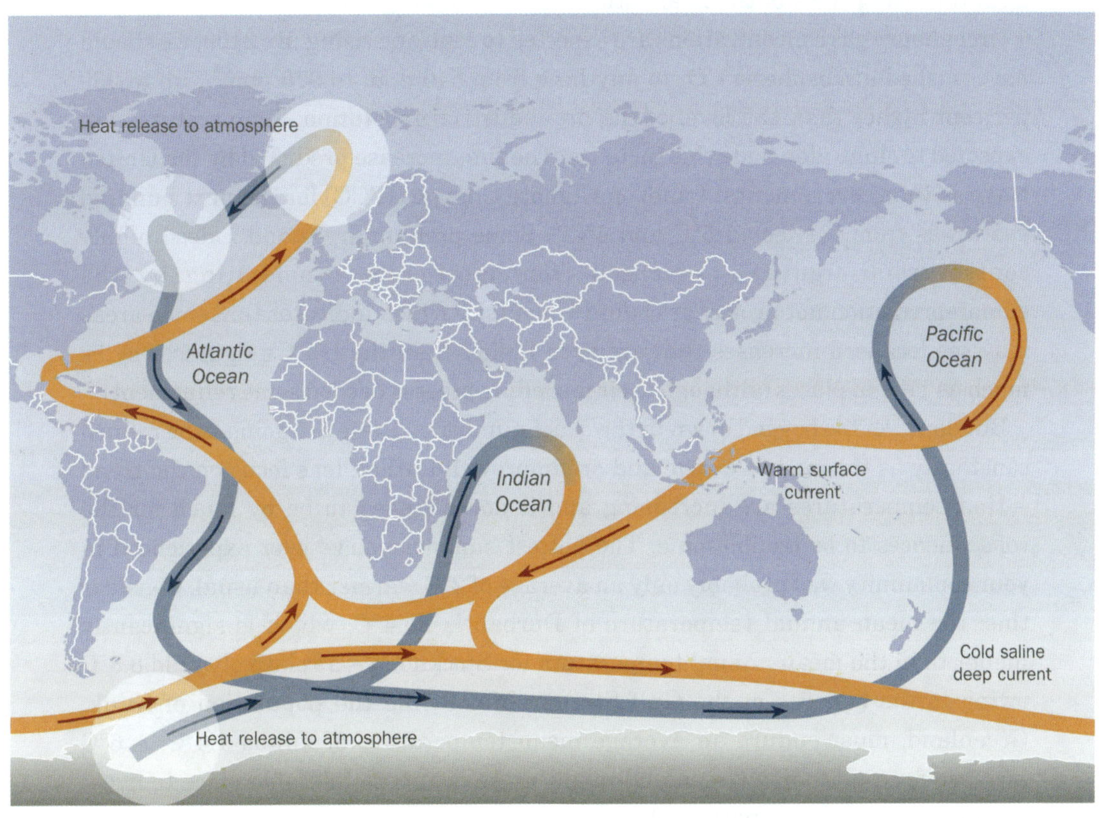


The global ocean conveyor belt distributes heat around the planet via enormous ocean currents. The Gulf Stream crosses the Atlantic from the Gulf of Mexico to Greenland, delivering warmth to the British Isles and continental Europe. Without the Gulf Stream, England would be 5°C colder on average.
Courtesy United Nations Intergovernmental Panel on Climate Change

the £50 million Thames Barrier – constructed to hold back tidal surges – is being considered for a £4 billion upgrade because it is not expected to save the capital from the rising tides of the near future.

Greenhouse gas concentrations are expected to continue rising. By 2100, we should have pushed atmospheric $CO_2$ to anything from 540 ppm to 970 ppm[28], up to 250 per cent higher than the level of the pre-industrial revolution. Temperatures are expected to climb alongside this increase. The first increase predicted by the United Nations Intergovernmental Panel on Climate Change (IPCC) for the next hundred years was from between 1.5°C and 6°C[29]. Some preliminary – and indeed highly controversial figures – emerged from more recent modelling by the climateprediction.net project at Oxford University which indicates that some areas may see localised increases that are even higher than the IPCC's figures – by as much as 11°C in places (although their modelling process needed some refinement)[30].

But don't get too bogged down in the exact numbers – these are going to be refined constantly as the models evolve and are improved. Rather, let's focus on the trend – that temperatures are increasing and it doesn't need to be by much for the consequences to be troublesome. The hottest summer you've ever experienced in your community was probably only an average of 2°C warmer than usual. Consider this: the mean annual temperature of Durban is 20.4°C, which is significantly higher than the mean annual temperature for Stockholm – a relatively mild 8.8°C owing to the benefits of the Gulf Stream. Meanwhile the population of Nuuk, Greenland, must endure an average mean temperature of minus 1.4°C. A 6°C increase for a place like Greenland might make it a more tolerable place to live – although it would be potentially catastrophic in terms of sea level rise and biodiversity change, but what about Durban? What about the Sahara, where the average mean temperature is only tolerable to the most hardened creatures?

Of course, it's not just the average increases that should cause us anxiety but the extreme or anomalous temperatures on the fringes of the average increase. Say, for instance, that Upington regularly experiences temperatures of around 40°C in summer but, increasingly, the record extremes reach 50°C or beyond. This would

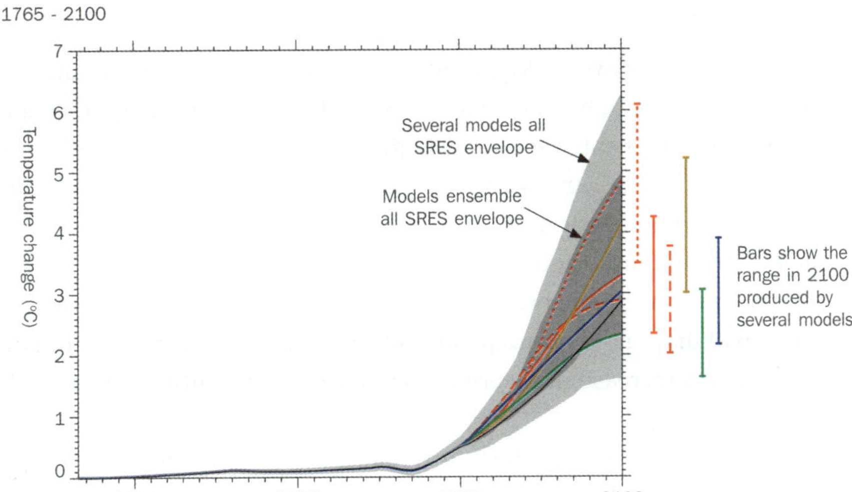



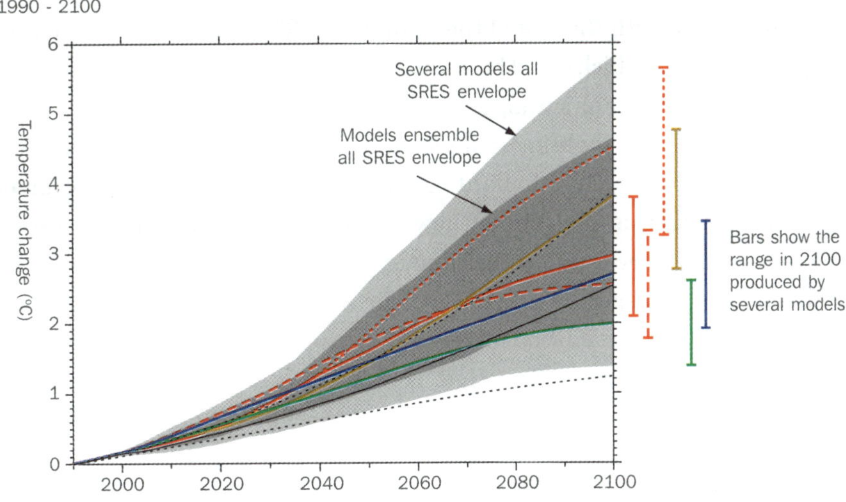

Temperature changes a) since 1765 and b) different modelled projections for temperature into the next century.
Courtesy United Nations Intergovernmental Panel on Climate Change

be devastating for livestock, the natural environment and, yes, young and old people who are less able to handle heat stress. Globally we are going to see more extreme weather events: heat waves and high temperatures; increasingly severe storms; some areas will become wetter or experience greater snowfalls – after all, in a warmer world, evaporation will increase and all that water vapour has to go somewhere because there is only so much the atmosphere can hold; other areas will become drier; temperatures will increase more over land, which means that places will dry out; more frequent and severe drought will occur in parts; tropical cyclones will increase in severity; and glaciers will continue their retreat.

• • • • • • •

Habitat is the space in which an organism lives. It refers to everything that a species uses in its home, for instance the amount of sunlight, food, the availability of shelter, the other species with which it shares the area, and even its predators. An ecological niche refers to how the species fits into that picture. It suggests an evolutionary moulding of a species by the factors in its habitat so eventually it fits hand in glove with the rest of the community. The total physical area that a species spreads across, which could be several habitats that differ slightly from one another, refers to its geographic range.

Climate is the single most important factor that ultimately decides whether a species can survive in a habitat or not. Here the concept of a bioclimatic envelope emerges, which is the crux of how scientists determine the future survival of a species in its home range as the climate changes. A bioclimatic envelope is an attempt to quantify an aspect of a species' niche. It is a mathematical summary of the full range of climatic conditions likely to be experienced by an organism over the full extent of its geographical range. How well that range is quantified and finally represented by its bioclimatic envelope depends on how much data is available about that species and its niche.

As conditions warm up, species are – in theory – expected to move towards the poles or higher in altitude to seek out their preferred temperatures. South Africa's plant regions, which are divided into seven broad categories – desert, forest, fynbos, grassland, Nama-karoo, savanna and succulent karoo – are expected to follow a similar 'migration' as they shift their distribution in response to changing conditions. The question always remains: how much viable habitat will they find as they shift their geographic range in response to a changing bioclimatic envelope?

To understand the impact of climate change on bioclimatic envelopes, scientists fed the extent and requirements of South Africa's different ecological regions into their number-crunching computers. These eventually spat out a map with a projection of the South African landscape by 2050 – the bioclimatic zones shifted south-east, with a warming and drying sweeping in from the north-west[31].

Within the next fifty to one hundred years:

> . . . the bioclimate of the country shows warming and aridification trends which are sufficient to shrink the area amenable to the country's biomes to between 38% and 55% of their current combined (regional) coverage. The largest losses occur in the western, central and northern parts of the country. These include the virtual complete loss or displacement of the existing succulent karoo biome . . . an extensive eastward shift of the Nama-karoo biome and contraction of the savanna biome to the northern borders of the country and its expansion in the grassland biome. The species rich fynbos biome . . . may . . . lose many species[32].

Within any habitat, a string of relationships will have evolved over millions of years: a flowering plant is pollinated by a single insect; a bat eats the insect; the plant's fruit makes a meal for a passing herbivore, which deposits nutrient-rich droppings on the ground for the plant before becoming a banquet for a pride of lion. These are the makings of intricate webs of life, food chains that keep nutrients circulating between organic and inorganic matter. We can never truly know how deeply the repercussions of climate change will be felt at every level within these close-knit communities if their specific requirements change. Most of today's plants and animals have evolved under global temperatures of about 3°C to 5°C cooler than the inter-glacial period in which we currently live[34], and we are changing

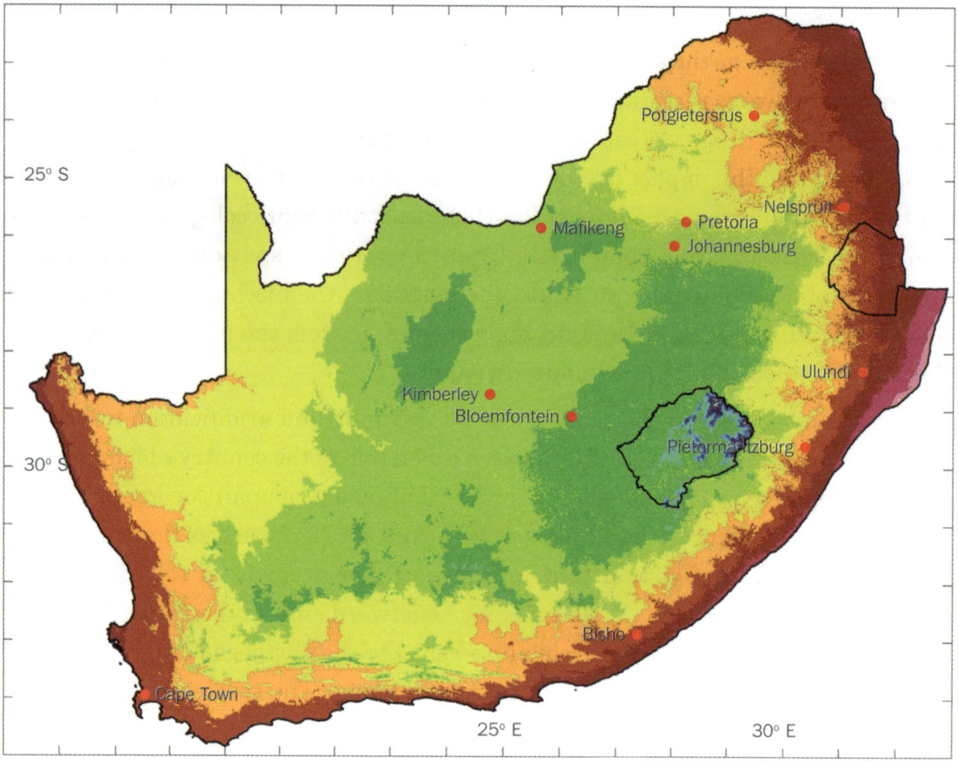

Current minimum July temperatures in South Africa.
Courtesy of the South African National Biodiversity Institute.

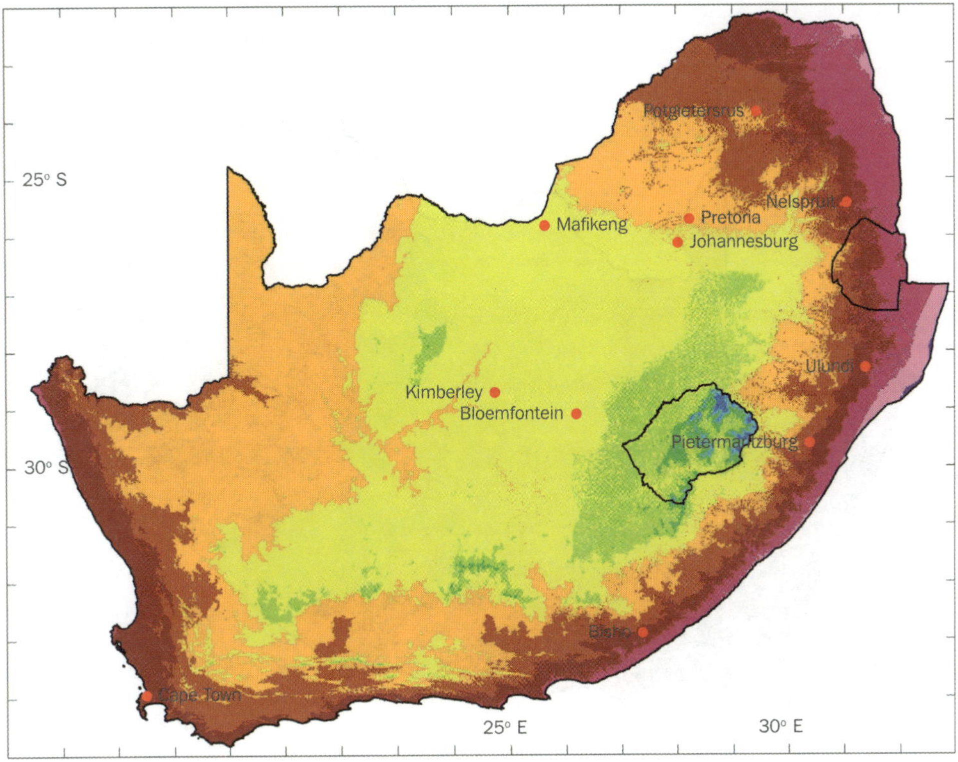

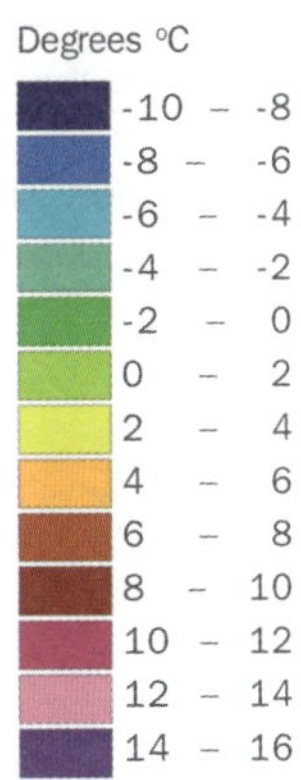


Projected minimum July temperatures for 2050 in South Africa.
Courtesy of the South African National Biodiversity Institute.

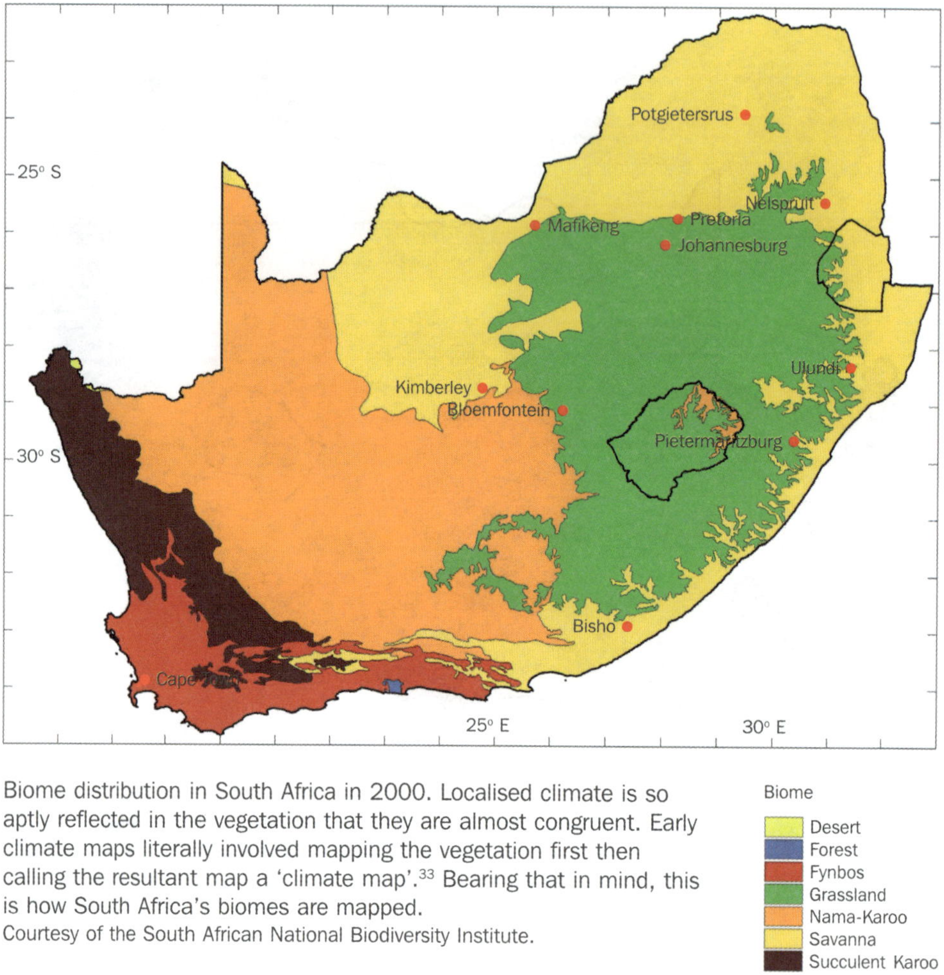

Biome distribution in South Africa in 2000. Localised climate is so aptly reflected in the vegetation that they are almost congruent. Early climate maps literally involved mapping the vegetation first then calling the resultant map a 'climate map'.[33] Bearing that in mind, this is how South Africa's biomes are mapped.
Courtesy of the South African National Biodiversity Institute.

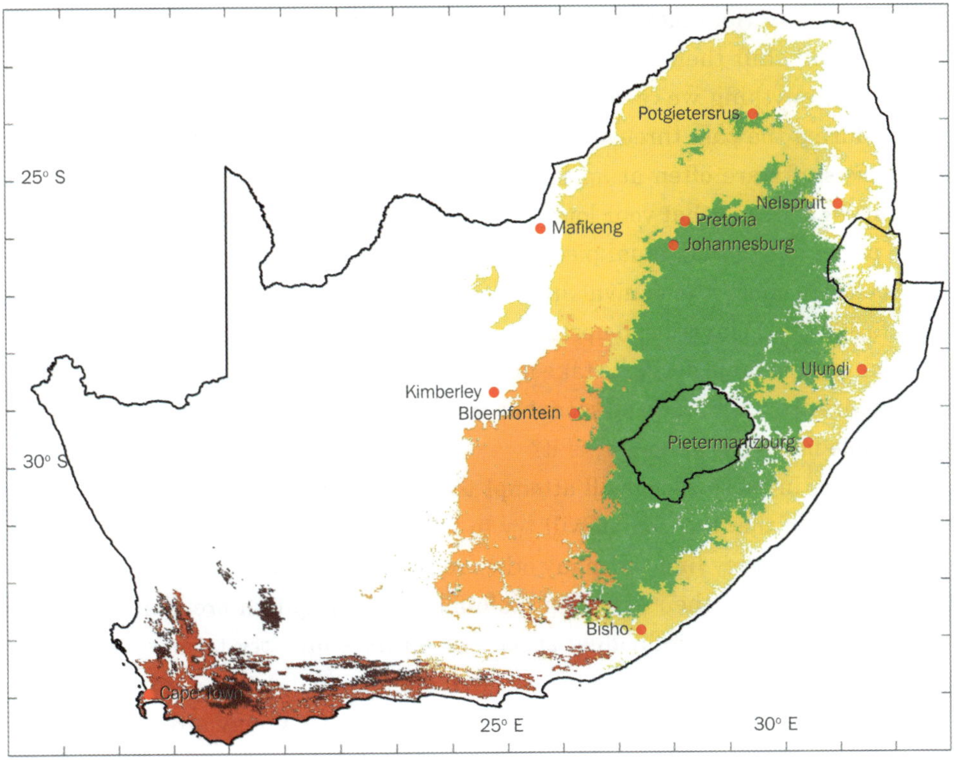

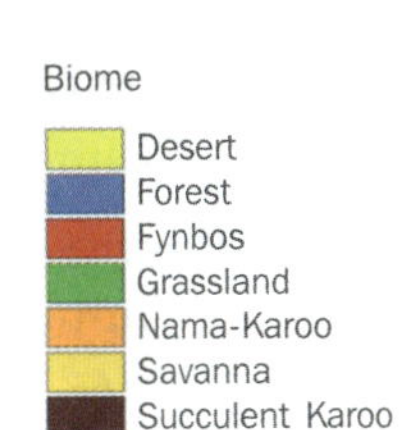



An illustrative map showing how biome distribution might look in South Africa by 2050, in response to an anticipated atmospheric carbon dioxide level of 550 parts per million. The areas left white will be novel landscapes, scientists can only speculate which species and communities will live here and which won't.
Courtesy of the South African National Biodiversity Institute.

conditions faster than most species will be able to adapt.

Species will shift their territories and, in some cases, species compositions will look unlike anything we've seen before. Many species have already shifted their ranges during the past three decades owing to changes in their localised climate[35] but these shifts are often at an individual level rather than a community level. Imagine for a second that your lounge decided to shift around at will. Not as a sum of the parts, but as individual pieces: for reasons entirely their own, the TV, lamp and bookshelf move the equivalent of several metres into the adjacent bedroom; the sofa and DVD player head upstairs into the master bedroom, while the speakers and dining room table decide to take up residence in the guest bathroom; the dining room chairs are nowhere to be seen, but might reappear in the garage. Not exactly a workable home environment, is it?

The plants in these zones will attempt to track their shifting climate envelopes and the rest of the food chains will try to follow. But when they get to their new destination, will they find the many other requirements that make a place home – the correct soil type, their attendant pollinators, the required prevailing winds or early morning fog? As individual species or entire suites try to move, they may bump up against obstacles – urban sprawl, oceans of wheat or maize, geological barriers such as mountains or impassable rivers. Or they may already be at the limit of their range – those species living on mountaintops that cannot climb any higher, or the fynbos species teetering on the southern edge of the continent. If the fynbos eco-climate shifts south-east, it will push out over the ocean where no terrestrial species can follow[36].

Adaptation to conditions at a genetic level differs between species but, in most cases, it takes millions of years or as many generations. Different species will respond uniquely: a common house fly, which might hurry through its lifecycle in 20-odd days, will produce many generations within a season, and hundreds of thousands within a decade. An elephant cow, which lives about 60 years, only becomes sexually mature at about 15 years of age and will have a calf about every four to five years. Compared to the house fly, the elephant is going to get through

far fewer generations in the same period of time. Which, do you think, is more
likely to adapt to climate ch___ at a genetic level?

___ms facing natural scientists – which species will
___ changes and which will not? Insects have been
___s' of climate change because of their ability to
___ging conditions. As for the other creatures that
___r many of them, their future is pretty certain.
___become extinct as their homes begin to unravel
___stribution of 1 103 species of plants and animals
___e predicted climate of 2050, 15 to 35 per cent of
___nction[37].

___rcise, but it highlights the urgency of the future
___has been documented – 51 species of butterflies
___titudes or, where possible, shifted their home
___ging of fishermen off these shores encountering
___y unrecorded hammerhead sharks, triggerfish
___hes. Octopus are being caught off the south
___reeding successfully in these same waters.
___undation (WWF), a '1°C rise in temperature
___n sole 200 to 400 miles north'[39]. In Antarctica,
___ne of only two ice-dependent penguin species,
___ of their range but declining seriously in the
___north. The sub-Antarctic system is creeping down into the former polar environment
of Biscoe Island near the western Antarctic Peninsula. Adelie penguins are
abandoning their breeding colonies here and are being replaced by gentoo penguins,
which are associated with the sub-Antarctic islands of the Falklands and Prince
Edward Islands[40].

Other changes are being documented:

Many plants in Europe flower about a week earlier than they did fifty years
ago and shed their leaves in the fall five days later. British birds breed an

average of nine days earlier than in the mid-20th century, and frogs mate up to seven weeks sooner. Tree swallows in North America migrate north in spring twelve days earlier than they did a quarter century ago. Red foxes in Canada are shifting their ranges hundreds of miles toward the pole, moving into the territories of Arctic foxes. Alpine plants are edging uphill and beginning to overrun rare species near mountain summits. [41]

There's a bias of scientific knowledge and weather records towards the Northern Hemisphere. Many in the south are playing catch-up as fewer weather records are available and less is known about ecosystem response to global warming. Nevertheless, the body of literature from the south grows each year and a picture is starting to emerge. Historically, it appears that during the past century Africa has warmed by 0.7°C with rainfall declining in the Sahel region but increasing in east central Africa, a trend that is going to continue into the future. The IPCC predicts that warming across the continent will range from 0.2°C to 0.5°C per decade, with greater warming occurring in the 'interior of semi-arid margins of the Sahara and central southern Africa'[42]. Predicting seasonal rainfall is a little more problematic: modelling of global change suggests that by 2100 rainfall will increase in the high latitudes and in some equatorial regions of the globe while a decrease will occur in many mid-latitude, subtropical and semi-arid regions, increasing water stress in the regions[43].

Locally, north and the interior of southern Africa may see a decrease in rainfall during the growing season by 2050 – by as much as 5 to 15 per cent in some cases. Climate modellers usually put different scenarios into their models – generally a conservative, extreme and moderate scenario, based on the future increase in concentration of greenhouse gases – and calculate the results. Obviously none of the conclusions will be identical, but the general consensus is that the east of the region will become wetter and the west will become drier. Remember the Mozambiquan woman who gave birth in a tree while escaping the floods in the region in March 2000? Incidences of flooding will continue in these parts, while an overall drying will sweep across the West and up towards the tropics.

The South African National Biodiversity Institute reported in 2001 that a general rise in temperature in South Africa would be experienced across the country, but with greater increases in the interior. The Northern Cape may see a 2.5°C to 4.5°C increase in January temperatures while the coast, which is moderated by the influence of the ocean, will see about 0.5°C to 1°C increase[44]. More recent and refined modelling has downscaled the results to be more region-specific. These results suggest that summertime changes will see temperature increases 'in excess of 1°C with the highest temperature changes in excess of 4°C'[45], anticipated during late spring and early summer. Heat stress and changed rainfall will probably result in greater runoff, evaporation and drying out of soils. South Africa 'will experience increased summer rainfall over the convective region of the central and eastern plateau and the Drakensberg . . . [while] the Western Cape will see . . . slight drying in summer and a slight decrease in wintertime frontal rainfall'[46]. Given that so much of the country varies from 'semi-arid to hyper-arid with only a few relatively humid parts where rainfall greatly exceed[s] 500 mm annually'[47], these changes could have a profound impact on the way the country looks in the future.

When scientists wanted to understand how the country's animals would respond to this change, they decided to try a theoretical simulation. They took the bioclimatic data for 179 birds, mammals, tortoises, lizards, snakes, beetles, butterflies, antlions and termites – all, to some degree, representative of animals across different family groups and different ecological regions in the country – and punched it into a computer climate model. Then they cranked up the heat by a theoretical 2°C and left the machine to crunch the numbers. What it spat out at the end looked something like this: all but three per cent of the species shifted their range in some way – 17 per cent of species expanded their range; 78 per cent contracted their range; two per cent became locally extinct. Most telling is that the bulk of the range shifts were in an easterly direction, which reflects the 'east-west aridity gradient' across the country. But some of the species in the east of the country shifted west, probably heading up onto the higher ground of the escarpment to escape the heat of the lower-lying areas. Predictably, species loss was highest in the west. Tellingly, 66 per cent of the species modelled in the Kruger National Park shifted their range to outside this flagship conservation area[48].

Troubled times are coming to the continent, both in South Africa and further north. The World Resources Institute's 2003 Millennium Ecosystem Assessment said that the collective problems of AIDS, population growth, urbanisation, disease, low literacy, poverty and political instability in regions of Africa would profoundly impact upon the cumulative effects and related pressures on biodiversity. Changes in land cover, spreading deserts and soil erosion, continued invasion by alien species, pollution and unsustainable use of resources[49] will continue to grind biodiversity into the ground.

About two years ago, on one of my first research forays for this book, I found myself facing my own troubled moment. It's a rather frivolous anecdote compared with the severity of the topic, but I'd spent the day driving through the near-desert in 40°C-plus without aircon looking for succulent plants in the shimmering quartz-covered Knersvlakte. I hadn't found any and, after a few hours of contemplating the rather bleak future of succulents in a changing climate, I was in need of a pick-me-up. You know the inveterate stereotype about women and shoes? Well, when I need retail therapy, I buy trees – which is a bit of a problem living in a third-storey flat. Nevertheless, that minor logistical hurdle did not hold me back when I stumbled upon the *Kokerboomkwekery* in Vanrhynsdorp up the West Coast.

To cheer myself up, I blew the equivalent of the cost of a refined Italian shoe import (a single shoe, not the pair) on a quiver tree. It's a youngster – probably the equivalent of a toddler in tree terms – and, at a stretch, its tallest leaves reach to my shoulder. I have planted it in a pot in my lounge, where the sun bakes it all year round. If the tree is cared for, it will outlive me three times over and grow to double my height. I'll have to provide for it in my will, along with the rest of my estate. Hopefully the person who takes it over when I am gone will know not to over-water it. The tree will probably outlive its second-generation keeper too, and be passed on again when he or she dies.

The thing about a quiver tree is that it does not buy into all that post-winter glitz of tropical trees. Where so many other plants become melodramatic with spring flushes and delicate green growth spurts, the quiver tree seems stoic and retiring – in keeping with the landscape from which it comes. But really it's just going about its business in the background, quietly pushing up one coronation of pale green after another. Little more than a handful of fleshy leaves emerge each year, slowly growing while no-one is looking.

Every day I ponder on that tree, that sliver of desert in my lounge which will live so many years beyond my own. And I contemplate the transience of the human species and its breath-like visit to this mortal coil. I look at the quiver tree and let it be a lesson to me.

# Chapter 3

The great, grey ghosts of the African savanna
Photograph © Don Pinnock

# Walking with the phantoms of Matabeleland[1]

*What is the purpose of the giant sequoia tree? The purpose of the giant sequoia tree is to provide shade for the tiny titmouse.*

Edward Abbey, naturalist and author, 1927–1989

It has been said that seeing a trumpeting elephant bearing down on you in full charge is like watching your gallows being built from the window of your prison cell. Not being able to see your executioner in mid-charge has to be exponentially worse. But, on reflection, it was quite worth the earth-stopping fear to consider – later – the extraordinary link between elephants,  the savanna system and climate change.

It was a glorious, warm winter's day on a termite mound in Zimbabwe's Hwange National Park when my heart faltered as a wall of trumpeting elephants descended through the ancient mopane veld. The thick, claustrophobic hedges closed in around us and smothered our view. I felt my breath catch in my throat.

Who knows what had spooked the animals, but they were reeling. Their panicked bellows cracked the sky. Five of us shared that thimble-sized mound but it was an extremely solitary moment. Just fifty metres away, to our left, the bush erupted as a mother and two youngsters broke through, running full tilt.

'Oi!'

Our guide, Steve Bolnick, bellowed out across the mopanes like a good Jewish lad.
'Oi, oi, oi, oi, oi!'

At first the young mother didn't hear him for her panic. Then, thankfully, she turned her weak gaze towards us, flicked out her trunk to taste the air and began peeling away.

It might have helped to know, in advance, that yelling like a banshee is a perfectly acceptable tactic for negotiating your way through a stampeding mob of elephants but there was no time for battle plans and strategising. You would have thought, too, that a herd of twenty-plus elephants bearing down on you at full speed would make the ground shudder. They didn't. The only sign that we were still in the direct path of the stampede was the advancing barrage of yelling, trumpeting, shrieking . . . any second we expected to see their great, round shapes break through the bush and engulf us.

It's a strange thing how time slows in moments of crisis. There seemed enough of it to check the most primal responses of my city-dulled instincts: I certainly did not have the tools to fight and, momentarily, bipedalism seemed to have deserted me. Someone else hadn't forgotten the instinct for flight (she was a triathlete, after all); she spun around and began to bolt, whimpering incoherently.

'Stay where you are! Don't move . . .'

I could not, even if I had tried. Still they came. Steve sent out another volley, a small signal that we were there and we really didn't want to cause them any further trouble.

'Oi, oi, oi, oi, oi!'

Like a weather vane spinning in an advancing front, the trumpeting turned and faded.

A piece of graffiti once told me that the problem with being an atheist is you haven't got anyone to thank when something really good happens. It had a point. Together we stood overlooking a chasm of silence. Not so much as a branch moved in the thicket to show how close we had come to confrontation with the largest land mammals on the planet.

Someone giggled. I think it may have been me.

Right there – in those few short, elongated moments – were as many lessons on elephants as you will learn in a studious day on the job. For instance, the claustrophobic mopane thicket in which we stood with its shoulders pressing in on us is typical of a certain type of elephant country. Because the trouble with these bulky herbivores is that they are big. To keep seven tonnes of body on the go, an adult elephant needs a fair bit of food – as much at 200 kg per day and almost as much as that in water. They trundle through the bush, picking up food as they go: leaves, bark, shrubs, fruit, grass. But they are messy feeders. They'll pull up tufts of grass which they don't really want, break off an entire branch to get a small trunkful of leaves, or strip off enough bark to cripple a tree. Elephants are even known to push over an entire tree just to show they can and then might nibble at the roots. This wasteful exhibitionism has given elephants a bad reputation in some circles, as though they are somehow irresponsible towards the environment. The truth is, elephants are the landscapers of the bushveld. While they may not be aware of it,   their role alongside a changing climate in how savannas are shaped in the next century is an intriguing topic for debate in scientific circles.

Another thing that became evident, there on that termite mound, is the precarious relationship of elephants with lions. Apart from humans, elephants don't have any predators – certainly not as they get large and unwieldy with age, even for the most lion-hearted meat eater. But in some parts of the continent, lions have figured out that if a mother elephant has two youngsters at her feet, she will invariably protect the younger of the two, leaving the older calf open to attack. It seems they have also learnt that the wonderfully useful prehensile nose is a fast track to downing a two-tonne adolescent. All a lion needs is to get a firm grip on the end of the trunk and hold on tight until the animal passes out, which sounds awfully like a blacker version of a Rudyard Kipling tale.

It might well have been lion that frightened our elephants. We had seen fresh spoor at dawn and the park's lion researchers later confirmed a large pride in the area. We had been told, as we started our first nervous steps in big game country, that elephants are among the four animals in the bush you don't want to confront, the others being lion, rhino and buffalo. It's not that elephants are aggressive or mean-spirited, they're just sensitive beasts with a lot of body weight to make their point.

A breeding herd is potentially the most dangerous because it comprises a gang of over-protective mothers with their youngsters. What you soon realise, after a day or two in the bush, is that besides the occasional lone bull or small bachelor herd, the only other kind of elephant you will encounter is one in a breeding herd. These mostly all-girl groups stick together under the nurturing trunk of their matriarch, which is usually the oldest and most experienced elephant and most likely related to all the younger animals in the group.

But they are not just out shopping for shoes. A recent situation at Hluhluwe-iMfolozi Park revealed social behaviour in elephants that is unnervingly like human nature. A herd of youngsters was relocated into the park. Without older bulls or a matriarch to fashion their social skills, they became delinquent. They frequently squared up with rhino, many times even to the death, and were extremely discourteous when encountering people or cars. Realising the importance of having mentors for the next generation, mature animals have since been introduced to iron out the social dysfunction among the youngsters.

Males leave the herd at sexual maturity but share the nurturing tendency – bachelor herds are also known to take in orphans. Elephant Camp, a commercial venture outside Victoria Falls, recently found a new baby mingling with its domesticated herd. Staff discovered that the baby had been delivered by a lone, wild bull which had dropped him off like an old traveller leaving a stray at the doors of an orphanage. He simply wandered in with the baby, introduced it to the herd, and wandered off again.

So close are the familial bonds between elephants – to the point where they even seem to recognise the bones of their dead – that culling no longer involves picking off one or two animals in a group. An entire herd has to be killed to save the remaining animals the trauma of loss.

• • • • • • •

Another day, another termite hill. We watched a small herd trundle past in single file a short distance away. Suddenly the leading lady picked up our smell on the breeze. The entire line came to a perfect halt. Then, as if to a military command, they turned on their left back foot and marched off in formation. It was too organised and precise to have been one elephant copying the body language of another.

For a long time, people thought there had to be something behind this almost telepathic behaviour. In the 1980s it was found that elephants can indeed communicate in a way we cannot detect. They emit deep, long-wave sounds, below the range of human hearing, which can travel as much as 4 km, if not further, when weather conditions permit. So while they appear to be silent, they are actually rumbling and purring to each other constantly and you will feel it, rather than hear it, if you ever get a chance to ride an elephant. It's like sitting on a contented cat. So when our ears registered the bellowing of stampeding elephants, they missed a far more colourful spectrum of sound as the animals veered away from us as one.

The feet of elephants are the latest body part to attract scientific interest. In spite of carrying up to seven tonnes around for sixty-odd years, their feet are incredibly sensitive. The sole is padded with a soft, pliable cartilage that cushions the foot with each step. Instead of the click-clack we are accustomed to from hoofed animals, an elephant's gigantic footstep is virtually undetectable. This is why elephants melt in and out of the bush with uncanny quiet, winning them the name in these parts of the great, grey ghosts.

Some scientists are even beginning to think that elephants may be able to use their sensitive soles to read their environment. Elephants have often been seen rolling unfamiliar objects gently under a front foot, as a blind person would finger Braille. A team of Stanford scientists are taking this a step further. They believe elephant infrasound may even be transmitted seismically through the ground and that the animals may be able to 'read' these sounds, transmitted to the ears through bones in the foot and toenail. This theory has yet to be tested, but scientists are broadcasting recorded infrasound seismically across elephant country in Namibia to see if the animals respond.

The strange thing about elephants is that one of their closest living relatives is the dassie (or rock hyrax). Standing an elephant next to a dassie is like putting a tub of margarine next to a double-decker bus. Nevertheless, they share enough features to suggest some kind of common ancestry: they both have an unusually long gestation period for their size – elephants are 22 to 24 months and dassies nine months. By comparison, a domestic cat's gestation period is generally around two months. Male elephants and male dassies both have internalised testes; the two species have similar construction of foot bones; and dassies have an extended tooth in the same position as an elephant's tusks.

The savanna of African folklore – with all the drama of its expansive grassy plains and scattered trees – is not some fixed place on the map, some lasting state of comfortable equilibrium. Rather, it's a transient state of an ever-shifting landscape which, over the millennia, has swung from straightforward grassland to congested forest and back again, along with natural changes in the climate. Today it stretches up the east coast of the country and along the north-easternmost borders where it joins the classic subtropical savannas further north.  These expanses once grazed a bewildering variety of animals, followed by the fastest and most cunning hunters on the planet. This is the Africa of the public imagination.

Savannas are paused in their current state of tree-to-grass ratio by a truckload of gardeners.  Seasonal rains keep savannas lush and green in summer. Wintertime drought stress keeps trees struggling to make it into maturity. Browsing by herbivores also keeps trees neatly trimmed back. Fires, rushing through the tissue-paper combustibility of grass, cut straight through tree seedlings like a voracious weedeater. Only when there are sufficiently long gaps between fires are young trees able to grow tall enough to withstand the next inferno. Thanks to fires, the animals of the savannas can enjoy the relative security of wide,   open spaces where they can see long distances and sprint either from danger or towards prey, depending on their position in the food chain. Without fire, particularly in higher rainfall areas, trees crowd in and a whole different suite of plants and animals take over[2]. The role of browsing and fire are accepted agents in the shaping of savannas.  But various other things have been found to work in concert with them too.

Professor William Bond of the University of Cape Town's botany department has been unearthing how trees work in the KwaZulu-Natal savanna. A slow and steady change has come over parts of the area in the past fifty years. Trees and thickets have crept over the landscape – thick, green and impenetrable. Farmers who have appropriated land for grazing have noticed it. Photographs taken in parts of Hluhluwe-iMfolozi Park in the 1940s show wide, open spaces and paths that today are congested with dense thickets. The financial implications are severe for both parties – loss of grazing for farmers and possible loss of tourism revenue for parks where their visitors can no longer view game.

Scientists such as Bond have looked at grazing and the management of the fire regime to explain this shift from savanna to woodland. When they looked at carbon dioxide ($CO_2$) concentrations present in the atmosphere, they made a startling discovery: an increase in atmospheric $CO_2$ could be 'fertilising' the trees, making them tick over like a fuel-injected car engine.

During daylight hours, trees 'breathe' in atmospheric carbon dioxide through the stomata (tiny holes in the surface of their leaves), at the same time losing water to the atmosphere from the moist interior of the leaves. Energy from sunlight is absorbed by the leaves, and water is replaced in the leaves once it has been transported from the roots by specialised veins. Through a chemical process in leaves (photosynthesis), $CO_2$ and water combine to form oxygen and carbohydrates such as sucrose. During daylight hours, oxygen is expelled through the stomata while $CO_2$ is absorbed to make carbohydrates, which are dissolved in the water and carried off to the rest of the plant. Here they are used to build the carbon-rich structure of the plant. This is the engine upon which the rest of life on this planet is completely dependent. Without photosynthesis, there would be no food chain, no breathable atmosphere and no daily grind from which you and I can escape at weekends. If you add more $CO_2$ to the mix, it makes sense that you are going to get more carbohydrates for the plant – add more sugar to the water bottle of an athlete and wait for the energy surge.

Bond and his colleagues at the South African National Biodiversity Institute (SANBI), Dr Guy Midgley and Barney Kgope,   put a few typical savanna trees into small chambers with a controlled atmosphere to see how they would respond to simulations of the different levels of $CO_2$ experienced on the planet since the most recent glacial period about 18 000 years ago. In the SANBI experiments, trees were allowed to grow in a $CO_2$ level of 180 parts per million (ppm) estimated for post-glacial conditions, in the 270 ppm of the pre-industrial era, and in present-day's 370 ppm. Some were grown in $CO_2$ concentrations as high as 550 ppm, anticipated for the year 2050, as well as in 750 ppm and 1 000 ppm.

Results showed that, under pre-industrial conditions, trees would have grown very slowly with less $CO_2$ available for building their carbon-rich architecture. However,  savanna grasses would have thrived as they don't require much carbon – 12 000 years ago, the eastern parts of South Africa would not have seen the typical browsing wildlife we associate with savannas today.   But as $CO_2$ levels in the experiments were increased, so too did the growth rate and robustness of the trees, both above and below the ground. Even when trees were clipped to simulate fire damage, the recovery rate was astonishing under simulations of current $CO_2$ conditions compared with those of 150 years ago.

It appears that $CO_2$ is a major factor driving the increasing abundance in tree saplings in savannas.   Where saplings would have died out as fires swept through the veld, they now appear to have enough carbohydrate reserves to re-sprout. Saplings are probably able to grow faster and to push quickly above the height where fire is lethal for them. Once they break through this 2 m to 4 m 'fire trap' and reach escape height, they have a good chance of reaching maturity. Grasses which thrived in the low $CO_2$ conditions of 12 000 years ago would find themselves backed up against the wall as the balance tipped in favour of super-fuelled trees. Higher $CO_2$ also means plants become slightly more water efficient and temperature tolerant[3].

Initially it would seem that a changing climate will benefit trees – and trees are supposed to mitigate against global warming by absorbing the $CO_2$ produced by burning fossil fuels. Surely this should be the answer to the problem of global climate change? It's not. Computer modelling has shown that, even if southern Africa were completely covered by trees, the amount of carbon absorbed from the atmosphere would be negligible in global terms[4].

The bioclimatic modelling in the *South African Country Study Report*, commissioned by the government in the late 1990s but only released to the public in a modified form in 2004, showed that the factors which shift savanna-type rangelands along the scale between forest and grassland will shift back and forth with global climate change. We have already seen how increasing atmospheric $CO_2$ could push savannas towards forestation. If minimum temperatures climb, the incidence of frost will decrease – frost has the same effect on saplings as fire, cutting them back and keeping the savannah stable. If frost reduces, saplings will have a greater chance of success.   Both these factors suggest that savannas will be shunted towards forest-type vegetation.

But throwing a spanner in the works is the trend towards aridification in savannas. Temperatures will climb by about 2.5°C to 3.5°C ,[5] with periodic drought stress probably gaining in severity, which will favour the success of grasses over trees. Anyone who has even given passing attention to how plants grow will have seen how they slow down in colder weather, wilt in very hot weather, and thrive at some point in between. Plot that on a graph and the result is hump-shaped[6]. The optimum temperature for grass growth is about 30°C whereas trees and shrubs are usually around 25°C. Growing season daytime temperatures in South African rangelands are already close to or above this limit[7],  states the *Country Study Report*. While it's unlikely that 'lethal maximum temperatures' for trees will be reached, the estimated 2°C increase will stunt their growth.

Increased evaporation and hence aridification under these conditions will shorten the growing season for trees if they are not made more water-efficient under increased $CO_2$. Basically, with more $CO_2$ available, the plant need not open its stomata as wide to breathe in, meaning less water is lost in the process and the tree then uses water more sparingly.

With all these factors pulling the savanna  back and forth, how can we know where it will be fifty years from now? Will it be grassland, woodland, or much as it is now? Dr Bob Scholes from the Council for Scientific and Industrial Research (CSIR), who has worked closely with Prof. William Bond on these studies, believes that the levels of $CO_2$ that are beneficial to plant growth have long been exceeded, probably at about 280 ppm[9]. Temperature and water are the two key factors for the future. A place like the Kruger National Park is already on the hot side of the temperature hump, meaning the region is heading away from optimal growth temperatures for trees and into dangerous territory.

Where do elephants fit into all of this? A good question – and it's one that the next generation of computer modelling is expected to answer. Where the *Country Study Report*  looked at temperature-water-$CO_2$-nitrogen interactions, Scholes believes that the next step is to include elephants into that equation. They may seem to be messy and wasteful consumers. But Prof. Norman Owen-Smith of the University of the Witwatersrand's Centre for African Ecology says elephants should not be seen in this 'narrow ecological perspective'[10]. Rather, they are as critical in the shaping of savannas as fire and frost.

Every plant-eating animal changes its environment somehow, but when it needs 200 kg of food a day, it takes things to a whole new level. Welcome to the world of the mega-herbivores[11]. Take elephants out of an ecosystem and it will get swamped with trees. Hluhluwe-iMfolozi Park lost its elephants in 1870[12]. Owen-Smith explains that fires kept glades open and expansive until grazing animal populations increased. More grazing meant less grass for fire; few fires meant more trees. Before long, trees were crowding in, extending their green canopies over formerly open spaces and creating thickets. The Kruger National Park saw a progressive loss of big trees following a steady increase in its elephant population as the animals began filtering back into the area from Mozambique after the park was established in the 1960s. The potential loss of habitat for vultures and high-nesting eagles may appear to be problematic.  But elephants also open up the savanna to accommodate a specific suite of species – once they clear away the trees, white rhino mow high grass down to a neat lawn. Shorter grasses then move in, as do the grazers. With their constant eating and excreting, a rich grazing lawn develops[13]. With fire kept out of this system, grazers become the recyclers of nutrients.

And when it comes to recycling nutrients, elephants are at the top of the food chain with their vast appetites[14]. Attractive as they are, trees hold onto their nutrients for the duration of their lives, except for the occasional shedding of fruit and leaves. Take the baobab – a tree with a 10 m diameter girth is probably 2 000 years old[15]. This means the sapling banked away its first nutrients when Jesus preached the sermon on the mount; the adolescent tree continued when the Prophet Muhammad warned Mecca against the perils of wealth and excess; it was still at it when Ghandi defied British Colonial rule in India; and it is still standing today. That's an impressive record by anyone's measure.

Every time an elephant pushes over a tree or breaks off a branch for a single mouthful of leaves, it is helping to recycle nutrients back through the system again. Like it or not, that's the natural order of things. For a long time scientists believed that there were only two possible stable states in the interaction between elephants and savannas:   grassland that is virtually denuded of trees or a woodland state. Now the thinking suggests that a third possible stable state lies somewhere in between the two – shrubby hedges where trees are pruned back by the foraging animals, allowing them to coppice into smaller, bushier shrubs. This was the kind of mopane veld we were walking through when we encountered our incensed elephants. Stunted trees like this offer leaves to any browser shorter than a giraffe, meaning there's a little more to go around for the average Joe of the bushveld.

This next step in computer modelling will help to explain how temperature, rainfall, seasonality, $CO_2$, soil texture, fire and mega-herbivores[16] might shape savannas in a changing climate.  Scholes already suspects that the effects of decreased water and increased temperature will negate any $CO_2$ fertilisation; that large numbers of elephant in a savanna will maintain a stable coppice state;   and that ultimately habitat alteration in a changing climate will depend on how fires and elephants are managed[17].

It's a frightful thing, finding yourself on the muzzle-end of a stampede, but it's probably good for sorting through any existential issues you might have. Feeling a little less frightened of the ghosts of Matabeleland (but no less respectful), we headed out into mountain acacia (*Brachystegia* woodland) in Chizarira National Park, east of Hwange, a few days after that fateful charge. Walking through blonde-bleached grass under a noon sun, aiming for the shade of trees where buffalo had flattened the grass like crop circles, our reverie was interrupted by a soft whistle from our guide. There, not 200 m from us, one elephant after another melted from the trees, browsing as they walked towards us. There was no panic, no trumpeting, no terror in the pit of my stomach. Just a quiet sort of nervy thrill.

We ducked behind the prostrate carcass of a tree and watched their approach, but quickly had to retreat into the shade of another tree. Despite their apparent dawdle, elephants cover ground quickly. There we sat, watching them doze under trees. One youngster experimented with balancing a branch on his head while another lost herself in a shower of dust. The matriarch – an enormous beast – meditated under a tree while her ladies picked through the foliage, looking for the best morsels. Theirs is a place where time ceases to exist, just the arrival and retreat of the Sun, as they move seamlessly though their world, feeding and nattering quietly as they go.

# Chapter 4

Fire, once the harbinger of life, now rips dangerously through fynbos
Photograph © Don Pinnock.

# Fire, fire!

*When one tugs at a single thing in nature, [one] finds it attached to the rest of the world.*

John Muir, naturalist, explorer, and writer, 1838–1914

It must have been a classic bushveld night. The dusk had settled under the eiderdown of night, a jackal yipped nearby and crickets serenaded the darkness. A campfire drew a circle of light around itself and tossed up showers of sparks. Shadows danced and wove to the tune of the flames. Braam Malherbe recalled being one of a clutch of adolescent boys around the fire, staring into its depths, hypnotised by the call of the night, the fire and lethargy after a day in the wilderness.

Their guide – a ranger – held up a boiled egg.

'Which part of this egg represents the Earth's biosphere?' he asked, breaking the silence.

No hesitation.

'The shell!' the boys chorused.

He cracked the egg with an affected, theatrical slowness, snapped away bits of shell and peeled off a strip of thin, transparent membrane.

Holding it up, he said: 'This is how fragile it is . . . and that's from the bottom of the sea to the top of our atmosphere.'

The memory fades and we are back on Signal Hill, sitting on a *stoep* overlooking Cape Town's sprawl. The chief executive officer of Cape Town's volunteer wildfire fighters sips his whisky and recalls the wilderness leadership course in 1974 when nature first imprinted itself on his consciousness.

'It had such a profound effect on me,' he muses.

Behind him the rim of the Atlantic Ocean is tipped with orange and blue in the wake of the day. Since that defining experience, Malherbe says he has felt a responsibility to the planet. Volunteer fire fighting in the Cape is how he gives back to nature and society.

Most summers, when the Cape's cold-season rains have long packed up their clouds and headed east, the notorious Cape Doctor sweeps fires into a swirling, snarling, biting, vicious fury across Table Mountain and the peninsula. In January 2000 nearly 8 500 hectares of fynbos and 17 homes were reduced to cinders by wildfires across the length of the Table Mountain range: in Deer Park above the City Bowl, Silvermine, Scarborough, and in and around the Cape Peninsula National Park (CPNP). Malherbe and a group of fellow volunteer South African National Parks honorary rangers offered their services to the section ranger for the Silvermine reserve and worked the fire non-stop for 32 hours.

Out of the ashes of this catastrophe, the Volunteer Wildfire Services grew. People from every level of Cape Town's social strata came forward in an extraordinary show of philanthropy – doctors, council workers, lawyers, secretaries, women and men – Christian, Muslim and atheists. They offered their time and, possibly even their lives, to saving the Cape's precious fynbos and many of the houses which dot the perimeter of the park from the devastation of these fires.

It hasn't always been like this. Fire was once the herald of life and renewal to these parts.

Botanists have carved up the world's plant kingdom into six distinct regions[1], where similarities in plant communities mean they can easily be grouped into a collective region. The Boreal Kingdom covers the high latitudes of the Northern Hemisphere; the Neotropic Kingdom covers Central and South America; the

Palaeotropical Kingdom covers most of Africa, the Middle East, India and Polynesia; and the Australasian and Antarctic Kingdoms encompass the continents and surrounds by the same names. These are all massive swathes of floral cloths overlaying wider continents.

The sixth region is the Cape Floral Kingdom. It is the smallest of all the botanical kingdoms and the only one to occur entirely on a single continent. It is swept into the bottom south-western edge of the African continent and covers an area as small as the size of Portugal. It has 8 600 species of plants of which 5 800 occur nowhere else in the world[2]. Spread over less than 0.04 per cent of the Earth's land surface, the Cape Floral Kingdom has 0.7 per cent of its endemic plant diversity. The Cape peninsula alone, which is smaller than the city of London, has more flowering plants than the entire British Isles[3].

This is a collection of plants so extraordinarily rare and unique that they are a wonder of the natural world. Crammed into the toe of the continent like a thin inner sole, the Cape Floral Kingdom forms a 200 km wide band running along about 1 000 km from Clanwilliam (just inland of the West Coast) to Humansdorp (near Port Elizabeth)[4]. The bulk of its residents are a collection of shrub-like plants known as fynbos, so named by early Dutch settlers who were not entirely enamoured with the *fijnbosch*[5] (literally 'fine bush') offered by their new home. Where the settlers needed constructive timber and firewood, instead they found the extravagantly frivolous embroidery of the Cape. Fynbos accounts for 80 per cent of the Cape Floral Kingdom's species, constituting four fifths of the total area, while the rest of the region is covered with small patches of succulent karoo, thicket, renosterveld and afromontane forest.

Some of the plants are so rare that they occur on a single plain on the Knersvlakte or hidden away in a little kloof somewhere on the Cape peninsula. In these cases, one poorly-considered housing development or a single block of Sauvignon Blanc vineyard could bring about a full-scale extinction. As a result of this high level of 'endemism', many of the Kingdom's plants are threatened with extinction, mostly in the face of spreading agriculture, housing and alien vegetation. More than 1 400 species are listed as critically rare, endangered or vulnerable; 29 species are already lost from the gene pool forever[6]; and a staggering three quarters of all the plants featured in the South Africa's Red Data book of threatened species occur in the Cape Floral Kingdom.

If this Kingdom were a haberdashery, then fynbos would be a priceless bolt of cloth. The sheer number and variety of its heath-like plants, protea shrubs, reeds and flowering bulb plants make a tight weave for one of the most richly unique plant communities on the planet[7]. But let me drop the 'spin' for a moment and be honest about fynbos. As a newcomer to the Cape, fynbos was about as breathtaking as a loaf of five-day-old bread. I've always found the acacia-studded expanses of bushveld or brooding montane forests far more appealing than the dreary scrubbiness of the Cape. Fynbos seems terribly drab at first, like a dull throw of mute green-grey, barely noticeable against the more majestic cliffs and peaks of the Cape Fold Mountains that it drapes.

Distance, I have since learned, does fynbos a disservice. To really appreciate the stuff, you need to be tripping over it. That's when you find an extraordinary weave of filigreed leaves, sprays of reeds and heady-scented flowers. Little shrubs made of the finest lacework, wiry and lean like a dancer. Frizzy pincushions; bell-like petals; riotously unrestrained lilies and orchids of the most delicate design; silver leaves shimmering in the wind and the strangely masculine protea flowers with their woody petals and bearded stamen. In the right season fynbos dances with eager pollinators. Game spotting is in miniature: bugs, birds, the finest reptiles and rare frogs.

This kingdom is no fairyland, though. It is one where summers are wracked by drought, winter squalls bring lashings of respite, the soils are left gaunt by a scarcity of nutrients, and fires regularly scythe their way across the landscape. Pretty as it is, fynbos is a graduate of the school of hard knocks. The soils in which it grows originate from mineral-deficient quartzite and sandstone rocks[8]. They do not have much in the way of nutrients to offer. On top of this, their coarse, granular texture allows water to seep straight through it, draining away nutrients from the bleached, leached, acidic soil.

As if that's not enough, the drought conditions that prevail through summer dry the soils to the bone. Vegetation must slow its growth to a crawl and use resources sparingly during the cooler mornings and afternoons, while taking a siesta through noon[9]. The summer gales that buffet the region for months on end only dry out the soil further, whipping away any humidity and desiccating the vegetation and leaf litter to a crisp. It's a mountainside of kindling just waiting for a spark. In such harsh conditions, combustion is inevitable.

It seems an unlikely place for such botanical splendour to settle and yet it's the very root of that finery.

Plants here have adapted to be extraordinarily efficient. They use these conditions to help them thrive and fire has been a keen instrument in the shaping of fynbos for millions of years. About 25 million years ago[10], long after Gondwanaland had split, the dinosaurs had been snuffed out and tectonic surges had pushed the massive Cape Fold Mountains into improbably fractured elevations, Africa was covered with tropical forests. In the Cape, rainforests and occasional swamps bristled with the ancient ancestors of yellowwoods, stinkwoods, olives, milkwoods, cedars and ferns. Conditions were warm, wet and tropical and, although the earliest signs of fynbos-type vegetation had emerged, trees reigned supreme.

The Antarctic continent finally sheared away from South America[11]. A cold ocean current began to circle the southern continent as it froze over, sucking up more of the planet's free-flowing water and causing sea levels to drop. Spinning off from this cold southern system came the frigid Benguela Current, which began its endless passage up the south-west coast of Africa, scouring the air of moisture and leaving the land arid and harsh. Tropical forests retreated, abandoning the Cape for the more hospitable climes of the east.

By five million years ago, the vegetation was morphing into shrubby grassland and woodland[12] but another climatic pattern was about to bed in around the Cape. A pressure system established itself here that shifts between two phases to this day: westerly winds bring rainfall during winter while relentless south-easterly winds drive away moisture and leave the place scorched during summer. The great retreat of the forests, brought on by the drying climate, was aided further by the fires introduced to the region by the new summertime conditions. Remnants of dried-out forests were reduced to embers. Finally, there came the birth of fynbos.

Without fire, fynbos becomes aged and barren. Before human settlement in the region, blazes easily raged across millions of hectares of fynbos. The lightning that accompanied the thunderstorms of late summer and early autumn would have struck the vegetation, left tinder-dry by the hot summer months when humidity was low and soil moisture depleted. This happened every four to 45 years or so, but averaged out at every twelve to fifteen years[13].

The charred and blackened landscapes left behind by fires give them a reputation for being destructive. But fires bring new life, like rain in the desert. Fynbos is honed by this adversity. Drought conditions and nutrient-poor soils have evolved these plants into small and twiggy runts. When fire sweeps through fynbos, the low fuel load produces a specific kind of fire – one that is mild and cool compared with the inferno associated with heavily-wooded forests. As they eat their way through the fynbos, flames split open seedpods, prune back proteas and release seeds buried by ants for safe keeping. From the ashes, seedlings sprout, shrubs send out new, fresh growth and bulb plants throw up magnificently coloured pillars of flowers.

Few fynbos plants are *not* triggered by fire to reproduce. More recently, it's been discovered that many seeds don't just need fire but the *smoke* from fire in order to switch off their dormancy. Heat from the fire breaks the seed casings and triggers growth – ethylene and ammonia in smoke stimulate some seeds to germinate. More specifically, the chemicals present in smoke derived from burned fynbos trigger the plant embryos to burst into growth[14]. (You can now buy a seed 'primer' from Kirstenbosch National Botanical Garden – disks of paper impregnated with 'fynbos-smoke-saturated water'. Simply add water and leave the seed to soak for 24 hours. The germination success rate on some species of fynbos almost doubles using this technique.)

Cooling autumnal nights, smoke taint and the first dashes of the seasonal rain start the germination of seeds that are adapted to respond to this specific three-part trigger. Vegetation is reduced to nutrient-dense ash and recycled into the soil[15], ready fertilizer for the next generation of plants. If it weren't for fire, fynbos would die out, allowing forests and thickets to creep back in.

But too much of a good thing comes at a price. Pine, gum and Australian acacias, brought to the Cape by settlers, are fast-growing, tall and rabidly thirsty trees that have muscled in on fynbos. They have changed fire's appetite. Not only does their growth, unchecked by climate or predators, trample roughshod over fynbos, but they provide exponentially more wood when the fires come. Hurtling across mountainsides and up gullies, the fires burn with a new ferocity. They are hotter, faster and more devastating than the survival mechanisms of fynbos are adapted to handle. Many shrubs are at a critically vulnerable stage after a fire. Seed banks are low and new shoots tender. If a fire sweeps through before they have had a few years to mature, it could be devastating.

Already one third of fynbos has been lost to agriculture, development or alien plants[16]. As if that's not enough, global warming is turning the region into a great powder keg. The temperature increase predicted for the area by computer climate modelling is still being refined, but an increase of about 1.5°C can be expected along the coast with a 2°C to 3°C increase occurring within the interior of the province's coastal mountains[17]. Pinning down the exact change in rainfall is not quite that easy, but the consensus points to a shift in the patterns of rainfall in the region. Winter rainfall will decrease, and will probably fall later in the season. Early modelling put this seasonal decline at 25 per cent. An increase in summertime humidity is expected to bring a little more rain in the late summer to the east and interior. (The intensity of these rains might well result in flooding events such as those seen in Robertson, Montagu and Swellendam in March 2003 when a cut-off low inundated the area with water[18].) The westerly storm tracks, responsible for the province's winter rainfall, are expected to shift southwards. This means that many of the fronts we need to bring the Cape's winter rainfall will head off out over the ocean, failing to bring the rain which, for millennia, has helped to drive evolution in these parts.

The Cape Fold Mountains are the region's rainmakers. As the left-over stumps of gnarled and twisted rock cut high into the face of seasonal winds, they force air upwards, cooling it suddenly and coaxing moisture from it. The modelling predicts an increase in this 'orographic' rain to the higher mountaintops[19]. The picture is still a bit fuzzy, but one thing is clear – the region is going to become hotter and drier with a greater intensity of extreme events such as drought, heat waves, sweeping berg winds and late-summer flooding.

As for the animals living in fynbos, most of the larger game – black rhino, gemsbok, mountain zebra – were hunted out of the region by the time my grandparents were born. But there's still a smorgasbord of little creatures that are as unique to the vegetation as fynbos is to the region. Take the Cape sugarbirds. They flit around the fynbos, snacking on insects and drinking nectar from the woody cups of the larger protea flowers. Hurrying from bush to bush, their heads dusted with pollen, they are among the great protea pollinators of the Cape. If they start shifting their home range in response to increasingly inhospitable temperatures, they could leave some proteas behind, unpollinated and ravaged by flames. Not only will sugarbirds be driven away from a food source that is critical to them during the breeding season, but protea species will begin to decline. Without the transfer of genetic material among shrubs, the vegetation could find itself in a genetic cul-de-sac. Of 300 species in the region, a staggering 50 per cent extinction rate is expected if these kinds of mutual arrangements collapse in this way[20].

Fynbos is packed with examples of co-dependency. The Table Mountain pride butterfly (*Aeropetes tulbaghia*), unlike most insects, is attracted to the colour red. It is the sole pollinator of at least 15 different red-flowered plants, including the orchid *Disa uniflora*. Some plants are adapted for pollination by bugs, others by rodents, others by birds. The examples are endless.

Here's an intriguing tale about a butterfly on the brink. It's called Dickson's sandveld copper (*Chrysoritis dicksoni*). It was discovered in 1946 near Melkbosstrand, just north of Cape Town, and it's believed once to have flown widely in undisturbed coastal fynbos in the Western Cape. This butterfly has long since disappeared from the Melkbosstrand location but was found a little further north, near Atlantis, and at a few other sites.

The curious thing about the sandveld copper is that, like some butterflies of its ilk, it is dependent on an ant for survival. But how this takes place is a fascinating saga of disguise, intoxication and deceit. First, the butterfly must make sure it is at least doing business in the same neighbourhood as the ants. To ensure this, the female butterfly lays her eggs on a specific plant near the nest of the Peringuey's cocktail ant (*Crematogaster peringueyi)*. The caterpillar knows instinctively that this highly mobile ant will always be in the vicinity of a particular scale insect (this relationship isn't understood yet but it has many lepidopterists curiously observing their behaviour). Since the scale insect is more sedentary, it's a safe bet that if the caterpillar hangs around long enough, the ant is sure to happen past. Once it hatches, the caterpillar will edge in close to the scale and wait for its ticket to the good life – right into the belly of the ant's nest.

No one is sure how the caterpillar gets there, but it probably follows the ant trail or is carried there by the ants themselves. The nest is a rugbyball-sized home constructed of a delicate papier mâché-like substance. It is built around the branches of low shrubs, about 30 cm above the ground, and is usually connected to a subterranean nest. Inside the nest, the caterpillar settles in, lines its new home with silk and waits to be fawned over by the ants. You may ask how the caterpillar gets away with such flagrant intrusion into so large, hungry and well populated a home. Simple. It barters with sugar.

The caterpillar has a tiny honey gland (called a dorsal nectary organ) near the tail end of its body from which it secretes a substance that casts a spell over the ants. Whether or not the substance contains a pheromone isn't known, but when it is released, ants in the vicinity appear intoxicated. They have been seen to stagger about drunkenly, as if on a high from something illegal. They cringe, heave as if to vomit, groom themselves and, if they wander off, may come back for more. When the ants are in this state of inebriation, the caterpillar raises the front of its body as if begging for food. It then locks mouthparts with one of the ants and feasts on food that the ant regurgitates from a crop in its abdomen. After feeding, the ant crawls over the caterpillar and hovers around the honey gland where it feasts on a sweet nectar emitted by the caterpillar. For ten whole months the caterpillar does nothing but lounge about and be waited on hand and foot, safely tucked away inside the bowels of the nest.

But the life of Riley only lasts for so long. Once nature compels it to change, the sandveld copper begins its metamorphosis into a butterfly. When it completes pupation, the new, soft-winged creature no longer has the magic honey gland to curry favour with the ant family and is completely unrecognisable to its former nannies. If it doesn't leave quickly, it will get eaten by the ants that nursed it to adulthood. It heads for the nearest exit, unfurls its wings and beats a hasty retreat.

It's an interesting symbiosis. The ant gets its sugar kick and the caterpillar gets life. Without the ant, the sandveld copper would cease to exist. Mysteriously, this particular ant occurs across the Western Cape, but the sandveld copper has only ever been found in a handful of sites near Cape Town since it was first identified. Its numbers are so scarce that it is listed as endangered and lepidopterists have recommended that these locations should not be published because one bout of over-enthusiastic collecting could reduce their numbers critically. One publication suggested that 'a diligent searcher could net a whole colony in a matter of minutes'[21].

This kind of localised home range typifies so much of fynbos and its attendant species and illustrates why a small disruption in the area of their limited flight path could wipe them out. I asked local butterfly specialist Dr André Claassens to introduce me to the sandveld copper. It turned out there wasn't much chance of that. It was mid-summer and this insect only flies between August and October. Furthermore, the site we wanted to visit had not produced any sign of the butterfly since 1995. It has probably become locally extinct. Nevertheless, we headed off to a place near Pella mission station, outside Atlantis, on a sweltering February day hoping that at least we would get a good idea of where it once lived. We might also see two other extremely rare butterflies – the Red Hill copper (*Aloeides egerides*) and the beautiful dark opal (*Chrysoritis nigricans nigricans*), which fly here too. Mostly, I hoped to catch a glimpse of the ant nest I had heard so much about.

After a few hours of scouring the neighbourhood, a few Red Hill coppers and one ant nest turned up in the search. More than anything, we found Port Jackson willow and plenty of litter. The population of Port Jackson seems to double every few years[22]. This, and the fires that follow it like hyenas after a kill, are probably responsible for the extinction of the sandveld copper from this location. Claassens and his peers have noticed a sharp decline in the size and number of ant nests over the years. They are convinced that the increased frequency of fire ripping through the area is responsible. It would make sense, after all. The ant nests are highly combustible and directly in the path of any fires which might sweep through here.

There is no scientific paper, supported by years of hard data, to show that global warming is categorically responsible for the demise of the sandveld copper. But we do know that this case study is a harbinger of things to come. Increased temperature and drought, along with the spread of alien plants, will bring ferocious fires through places just like this one and throw out of kilter the delicate systems that have taken many millions of years to evolve[23]. Whatever flicked the switch for the sandveld copper, we could well see many more creatures follow the same path in years to come.

Seventy butterfly species fly on the Cape peninsula, many of which do not occur beyond the Western Cape Province. Three of the peninsula's butterflies occur nowhere else but here: the peninsula skollie (*Thestor yildizae* [Kocak]) flies on several of the peninsula's mountains; the Lion's Head copper (*Trimenia malagrida malagrida* [Wallengren]) has not been seen in its last known locality near Cape Town for about ten years and may be extinct, perhaps partly due to repeated mountain fires occurring here in quick succession; and the barbers ranger (*Kedestes barbarae bunta* [Evans]) is only known to fly from the peninsula to the Cape Flats (around Steenberg, Retreat and Strandfontein), but its future is threatened by habitat destruction and the spread of alien vegetation. So much for our pretty winged things.

Fires aside, the shape of things to come in a changing climate is already pretty dismal. The next 50 to 100 years could see mountain fynbos unable to retreat any further up slopes to cool refugia, and pockets of rare species could be pushed to the brink. Thirstier species may expire as droughts become more frequent and severe.

• • • • • • •

Using five parameters that are imperative to the growth and survival of plants within the fynbos biome, scientists used the theoretical approach of bioclimatic modelling to see what the map of the fynbos biome would look like 50 years from now. The conclusion was that a 50 per cent loss in the fynbos biome envelope could be expected by 2050, with the greatest loss taking place in the northern limits of the region where less than ten per cent of the fynbos community would be able to hold out above 33° south[24]. Species growing high on mountain slopes or towards the southern reaches could also be most vulnerable. When the climate-habitat window shifts south-eastwards, pushing the biome off the southern tip of the continent[25], these species would have nowhere else to go but into the grave. Since the biome has a high proportion of endemic species, any loss of range would result in extinctions.

Increased fire activity is only going to make a bad situation even worse. A large number of the more slow-growing, slow-maturing plants, such as many of the proteas, are expected to suffer most under the onslaught. They will be stunted by the prevailing droughts and cut back more by fire. 'If the frequencies [of fires] are reduced from once in 15 years to once in 5 years, many plant species that are killed by fire, and that rely on re-seeding to survive, would become extinct as they would not have enough time to mature and set seeds between fires[26].' The years between fires will become fewer, as will the recovery time for the plants here. In many cases this changing fire regime has already resulted in extinctions in the protea community. Above the city bowl, a type of pincushion protea (*Leucospermum vestitum*) is lost to Lion's Head[27] and the Burchell's sugarbush (*Protea burchellii*) is no longer found at its home on Signal's Hill and Devil's Peak due to overly frequent fires[28]. Loss of vegetative cover then impacts on the runoff of rainwater, the stability of soil and the rate of erosion.

The Cape Peninsula National Park volunteer wildfire fighters are doing their utmost to save biodiversity from the ravages of flames. The volunteers have grown to such a size and efficiency that Cape Nature Conservation has called them to assist with fires beyond their jurisdiction – into the Hottentots Holland mountains, the Koegelberg, Cedarberg and surrounds.

Glamorous the volunteer service may seem to new recruits who turn up at the Newlands forestry station to sign onto the team, but training requires a hard slog all year round. In regular training drills recruits go through basic hose and pump exercises, learn to operate the necessary machinery and engines, master abseiling down mountains and rappelling from helicopters as well as on-the-ground fire-fighting techniques. During fire season they give up every second Saturday to practise evacuation procedures, usually involving a volunteer strapped into a stretcher who is abseiled down the mountainside. When they are not doing that, they assist with fundraising and clearing the park of alien vegetation.

A wildfire can be a dangerous beast – something many of the initiates discover with time. One of Malherbe's colleagues learned the hard way when a fire in Deer Park caught him unawares.

'The ranger was in his vehicle when the fire swirled up a gully,' recalls Malherbe. 'On an incline, a fire can move many times faster, depending on the gradient and wind. The gully created a vacuum and, because it was a drainage area, it was filled with thicker vegetation.'

The flames swept over the vehicle.

'He must have panicked because he got out of the vehicle . . . you never get out of your vehicle. The smoke was very thick, I couldn't see a thing. I just knew he had gone in there.'

Suddenly the ranger staggered from the smoke. He wasn't wearing the standard fire-retardant gear that day because he hadn't come prepared for frontline fighting. When he collapsed into his colleague's arms, the skin on his arms slid off like butter. He was wracked with spasms of shock before he reached the Cape Town Medi-clinic a short while later, where he spent months recovering and undergoing skin grafts.

'He's just getting the use of his hand back,' says Malherbe solemnly, using one hand to push the other into a stiff fist to demonstrate the ranger's recovering limb.

Recounting this tale on a cool winter's evening, Malherbe ponders the human condition.

'If we carry on destroying things, we lose the animals, amphibians and birds . . . what's left? Maybe I'm naïve and utopian but we have a brain. We should be learning from all of this . . . from the past.'

He pauses, cocking his head slightly, and listens for a moment to a chirrup from the shrubs nearby.

'The robins are starting.'

Suddenly all hints of the past day wash away into a black-blue ocean and the thin membrane of life is overcome by night. The stationary luminosity of a verandah light enshrouds us and our shadows are mechanically still. Something's missing, this close to the city. It must be the call of a black-backed jackal.

# Chapter 5

Another kind of Baggins, Bilbo's rain frog
Photograph © Les Minter

# Love songs in the mist

*Every dewdrop and raindrop had a whole heaven within it.*
<br>Henry Longfellow, poet, 1807–1882

Bilbo Baggins has a doppelgänger outside of the Shire. On misty days he can be heard whistling across the rolling hills of the KwaZulu-Natal Midlands. Just like any hobbit, this fellow is rare and shy but, if you are lucky enough to come across him one day, you will see the resemblance. He's stout around the middle and tends to waddle a bit when he walks. His nose is stubby, his eyes forward-looking and he has a quaintly receding chin. He's a little more grumpy-looking than your average hobbit (the down-turned mouth doesn't help) but then you can't expect to be cheerful every day. Only this Baggins is not a hobbit. He's a frog.

The discovery of this frog in its grassy Midlands hideaway was entirely serendipitous[1]. A mini-bus was winding through Babanango about ten years ago when a high-pitched, rattling whistle cut through the sound of the engine. Fortunately the driver was familiar with the language of rain frogs and immediately recognised an unusual dialect. It seemed to have a rapid-fire delivery rather than the usual drawl of your average rain frog call. Dr Les Minter stopped the bus and began grubbing around in the undergrowth until he found an entirely new and undiscovered rain frog. Today it is known as Bilbo's rain frog (*Breviceps bagginsi*), named after the hero in Tolkien's book *The Hobbit* that Minter once read to his children.

Not much is known about Bilbo's rain frog – science has only officially been aware of it for three years – because most of its natural habitat has been lost to pine and gum plantations. Today it is only found buried in the sandy soils of grassy road verges, which are probably the only places that resemble its original home. When it rains, Bilbo's rain frogs come out to sing. If you know what to listen for, you might pick up its song on the air somewhere on the roadside between Babanango and Melmoth, or near Howick[2]. If you pay attention, you may even see him out and about when the termites fly, snacking on these peanut-like delights.

My own search for Baggins and his ilk started with a subtle clue from nature twenty years ago. It's a mood rather than a precise event now, frayed as it has become with time and mental fingering, but it happened something like this: I remember travelling through the hot, grassy Thyume Valley near the town of Alice in the Eastern Cape. Our old VW Combi clattered past a meditative huddle of Nguni cattle as they re-chewed lunch. We made our way round a switchback and began the climb. Before long we came upon the regiments of trees that march down the pass, bending down over us, layer upon layer of leaves and monkey vines dampening the air.

Hogsback is all about the forests, as you will know if you have ever visited the Amatola Mountains. Aging yellowwoods with girths hiding centuries of growth-rings; samango monkeys lolling about on branches; a lone bush pig rummaging through the undergrowth; cicadas ticking in the background like a high-speed metronome; marauding Cape parrots and scarlet-winged louries flashing outrageously through the canopy, raising hell as they clatter about in search of nuts and berries. Humid and hot, the air grips the afternoon in a lethargic reverie. Then suddenly, as if with a massive sigh, the forest breathes out ribbons of mist that curl up from its bottle-green roof, tumble together and roll up the mountainside. The muggy day gives way to another velvety front. The mercury drops and the finest sprinkling of drizzle settles like silver jewels where it falls.

From somewhere deep inside the white cloud tinkle the first chords of seduction: soprano whistles, throaty croaks, woody rattles, staccato chirrups and trills, twinkling out into the air like fairy lights on a Christmas tree. Right there, in that moment, is where the love songs begin.

Standing barefoot in the mist, I would listen discreetly to the love incantations of amorous frogs. Just then it would begin to rain.

It has only occurred to me now, after delving into the lives of the characters in this story, that much of my own life has been spent within singing distance of some of the country's most threatened and endangered amphibians. Hogsback has two of them. The Hogsback frog (*Anhydrophryne rattrayi*), or Rattray's forest frog, was first discovered in Hogsback by a school principal and has since been found at a few sites within the afromontane forests that hug the mountain slopes of the region. Its contribution to my listening pleasure would have been the melodious *ping, ping, ping* of the male calling for a mate. These tiny creatures – the female measures in at about 2 cm in length – reside in the leaf litter of the forest floor and dig nests for their eggs in the clay soil beneath it[3]. Most frogs need water to breed, which is why they break out in song when the rains come – something like a 'hurry, hurry, let's get to it before the weather clears!' The Hogsback frog is a little different. The tadpole is no water baby and would drown if it were dropped into a pond. It needs just enough moisture in its earthy home to grow from egg to tadpole to adult.

Twenty years after I first heard it call, the Hogsback frog is listed as endangered because its habitat has been shaved away by commercial forestry and the steady felling of natural timber. Their communities have always been remote from each other, like villages scattered across a vast countryside. This makes them extremely vulnerable to any changes in their environment.

If the forests here were to shift uphill, in keeping with the trend to escape climbing temperatures, would the Hogsback frog be able to move with them? Those forests that still have hills available to move up would probably find their migration cut off by farmlands, alien tree plantations, buildings and roads.

Another of Hogsback's endangered residents is the Amatola toad (*Bufo amatolicus*). Where the Hogsback frog settled into forests just above 1 100m in altitude, the toad prefers the moist grasslands higher up the mountains. But it has only been found at a few places on the Amatola and Winterberg mountains, Hogsback being one of them. Its nasal squawk must once have been heard across kilometres of grassland on misty days before pine plantations spread out across the region. Now their grassy ponds are in a standoff with altered fire regimes and the sinking water table associated with these heat-burning, liquid-sapping trees.

Frogs have a curious adaptation in that they can breathe through their skin. They do have lungs, but their skin can draw in oxygen, expel carbon dioxide and is permeable to water. It's a delicate epidermis without the protection of hair or scales worn by so many other animals and it needs to be constantly moist. For this reason, frogs are the canaries in the mineshafts of a changing environment. Their delicate skin makes them tremendously sensitive to the kinds of changes that humans bring to the environment: chemicals leaching into the ground, pollution, the drying out of wetlands. There is little honour in being the community's indicator species.

It is sobering to realise that the activities of a species as young as ours could so easily annihilate a group of animals that has outlived even the dinosaurs.

The story of humans goes something like this: a group of reptiles called therapsids evolved into mammals, the warm-blooded, mostly hairy animals from which humans descended. This happened at about the time when dinosaurs – also reptiles – were rising up to dominate the planet during the Triassic Period about 195 million years ago. Mammals had it tough at first, struggling against the superiority of the dinosaurs. But the planet's climate swung into a cooling trend during the Mesozoic Era. Cool-blooded dinosaurs, which leached their heat from the environment, became sluggish and faltered as the temperature swung between greater extremes. The dinosaurs, many scientists believe, were on their way out when that meteorite struck, obliterating them completely. Any mammals to survive the carnage limped out of settling clouds of ash and grew up to become modern mammals[4].

Modern humans are a recent offshoot of that evolutionary tree, coming along just yesterday in geological time: about 200 000 years ago. The story of frogs goes back a great deal further. Long before the dinosaurs, the first amphibians crawled out of the planet's waters and took up life on land. A segue or uninterrupted transition between aquatic fish and land-based reptiles, their early crawling bodies represented a watershed in the evolutionary chain. About 370 million years ago[5], 'fleshy-finned fish'[6] – probably trapped in drying ponds during droughts – found themselves becoming less dependent on water, eventually evolving lungs and limbs. These creatures were the first vertebrates to settle on land and adapted marvellously, evolving nature's first eyelids, sense of smell and extendable tongue[7]. Some of these creatures would later go on to become reptiles, where their tough skin and shelled eggs gave them the upper hand in the tough arena of survival[8]. Some of these scaly creatures evolved into the mammals which later produced humans. But various amphibians continued on, becoming today's croaky little lovers.

Testimony to this long ancestry lives not far from my current home in Cape Town and its chance of survival is even more compromised than Hogsback's species. The Table Mountain ghost frog (*Heleophryne rosei*) is a member of a family whose immediate relatives date back 200 million years to the time before the supercontinent Gondwanaland began splitting up[9]. It is adapted to fast-flowing streams in fynbos and its tadpoles are like pool-cleaning Kreepy Krawlies with large sucker mouths to help them cling to rocks in the rushing water[10]. The Table Mountain ghost frog breeds in exactly six streams, most of which run down the wetter south- and east-facing sides of the mountain. That's it. Nowhere else.

Consider, for a minute, how much of the planet is covered by humans and our farmlands. Think about how much tarmac and concrete have come between earth and air as a dry, lifeless and impenetrable artificial shell. Now contrast this with the six small water courses available to this frog. Fast-flowing, clear water is its sole habitat so, where streams have been dammed up, this water becomes a stagnant desert for them. 'Save the Tahr' campaigners would do well to consider that the presence of this exotic goat on Table Mountain may have eroded stream banks and muddied the frog's few water courses as they overgrazed the mountain and trampled streams.

Like so many frogs in the Cape Fold Mountains, this one has adapted itself to a specific set of conditions: its tiny home range, rapidly flowing streams and the winter rainfall peculiar to this part of the country. Because it has found its place in cold, nutrient-deficient water, the Table Mountain ghost frog's life cycle is slow compared with many warm climate frogs, taking over a year to make it through tadpole-hood[11]. This means that a string of dry seasons could be fatal to the next generations of these frogs. The warming and drying expected in the Cape in the next few decades will move this frog, like so many other rare species in the fynbos biome, even closer to the abyss of extinction. Already it is labelled as critically endangered. For the Table Mountain ghost frog, it's a bit like being a farrier at the time when Henry Ford put horses out to pasture.

I wanted to find out more about frogs in a changing world, so I tracked down conservation biologists Atherton de Villiers and Andrew Turner from Cape Nature's scientific services and asked them to take me frogging. These biologists are in the early stages of a long-term study, monitoring two sites high in the Hottentots Holland mountains. Six times a year, they go out counting frog songs – mostly in winter and early spring when these animals call. We visited a site about 1 000 m above sea level in a spongy mountain seep that towers behind Somerset West. It is the source of the Riviersonderend River, where water trickles out of the sponge, down the mountainside and joins other such streams until it becomes the main river that spills out into the Atlantic at the Breede River mouth.

The chance of locating a frog in mountain fynbos imitates the whole needle-haystack scenario because frogs prefer habitats that are dense with reed-like restios and mosses. The easiest way of judging the health of a frog population is to look for signs of breeding and all this requires is a good pair of ears. Together the froggers stood, hands cupped behind their ears like makeshift satellite receivers, staring meditatively over the mountainscape. They were training their ears to isolate the possible calls of twelve different kinds of frogs known to live up here. To me it seemed about as easy as picking out a faint cymbal strike from an orchestra but, between them, they counted three different frog songs. Rattling, humming and trilling from the reedy water seep came the advertising calls of the De Villiers moss frog (*Arthroleptella villiersi*), the Cape mountain rain frog (*Breviceps montanus*) and the banded stream frog (*Strongylopus bonaespei*).

Once the biologists had satisfied themselves with enough aural data, it was time to search for a frog that no one will hear – the Cape mountain toadlet (*Capensibufo rosei*). This curious animal is the only voiceless frog or toad in the whole of southern Africa – probably related to the fact that it doesn't have ears. Where other frogs call out to potential mates, these toadlets head *en masse* to a specific breeding pool – they seem to be loyal to specific pools – and indulge in an orgy of breeding. Turner found a toadlet after carefully lifting rocks along the edge of the seep. Most adults apparently reach about 3 cm to 4 cm[12] but this one was tiny – not much longer than the trimmed nail on my index finger – and glistening blackly. Annoyed by our intrusion, it puffed itself up and put everything it had into being intimidating. Sitting in Turner's hand it looked like a highly angry and minute blimp. More delicate a creature there surely could not be.

Ironically, natural to-and-fro swings in climate are probably responsible for enticing these creatures up here in the first place, and it's how they became specialists in high altitude mountain survival. Think about it this way: there are numerous species of frogs spread out over various parts of the country, with the densest populations forming two common groups – one in the Western Cape and another up the east coast and eastern interior. As climate vacillated between cold glacial periods and warm interglacial periods, as it has over the past four million years[13], the frogs' home ranges swept back and forth in time to the natural rhythm. Some frog specialists have shown that the distribution of the Western Cape species would have expanded in cooler times, while the more tropical species would have retreated, with the reverse happening in warmer times[14].

As these species migrated back and forth, tracking the climate, some became caught up in geographical nets such as mountain ridges or islands of land. Trapped in these refugia and isolated from other communities, these small pockets of frogs drifted from their genetic blueprint if they were given enough time. As long as the current of change that swept them there was not too dramatic to kill them off, all they needed was a community big enough to prevent them from being suffocated in the crush of a genetic cul-de-sac but not too large to overcome the genetic drift. If all the ingredients came together, a new species would inevitably occur[15]. This explains the distribution of different frog species across the Western Cape and also speaks of the genetic uniqueness of the plants in the Cape fynbos and the Karoo's succulent communities of the West Coast.

Many of the frogs dotted about on the upper reaches of the Cape Fold Mountains are too geographically removed from one another to be able to move between peaks. Yet many are genetically related. This probably means that they were once part of a single, larger community that must have settled further down the mountains. Conditions would have been cooler and wetter, allowing the frogs to spread lower down the mountains and across a wider range, and to merge into a larger community or move between less fragmented ones. As the climate warmed and dried, they retreated uphill into lofty eires where they have stayed ever since.

The problem with global warming is that when change comes, it will probably be too fast for species to adapt at a genetic level. Many high altitude frogs and toads could be at the ceiling of their possible range shift – they have no more 'up' to which to move. For the frogs living in this mossy, cold little world, things could become a lot drier and more troublesome in the next fifty years.

Another threat also comes from the sky. The depletion of atmospheric ozone in recent decades, largely through CFCs (chlorofluorocarbons), means that greater amounts of UV-B radiation reach the planet. We have already seen increases in skin cancers in people as a result and frogs don't have it much better. UV-B can scramble their DNA, hammering their eggs and tadpoles, generally eroding their breeding success[16] and spawning weird genetic offshoots. There is some good news for fynbos frogs. The mountain waters in this area are full of nutrient humus, tinted red-brown from tannins and vegetation to look like well-stewed tea. The colour might act as a natural sunscreen, filtering UV the way that dust does in the atmosphere.

A global decline of high altitude amphibians in pristine environments has been linked with atmospheric pollution and climatic change[17]. The *Atlas and Red Data Book of the Frogs of South Africa, Lesotho and Swaziland* states:

> These factors may lower amphibians' resistance to disease. There is no evidence, at present, that global factors have caused declines in southern African amphibian populations, but such threats may be significant.

Globally, frog numbers have dropped radically. The causes? Our appetite for frogs' legs and pretty pets (it's amusing, in a voyeuristic sort of way, that some people eat the animals that others hold dear as pets); the spread of infectious diseases; and non-indigenous predators. Locally we are changing fire regimes, particularly in fynbos and grass systems. We are introducing alien plants into delicate ecosystems; ploughing up land for agriculture; polluting water courses with fertilizer run-off, herbicides, pesticides, petroleum products and heavy metals. We are carving up and degrading the habitats of frogs, constructing dams, driving cattle into moist grasslands, and building roads across their breeding migration routes[18].

The *Frog Atlas* singles out species in the succulent karoo and northern reaches of fynbos as those most likely to suffer in the warming and drying that will be most severe in these regions. Should you like to acquaint yourself with the Namaqua stream frog (*Strongylopus springbokensis*), the desert rain frog (*Breviceps macrops*), the paradise toad (*Bufo robinsoni*), the Namaqua dainty frog (*Cacosternum namaquense*) or the Karoo dainty frog (*Cacosternum karooicum*)[19], you would be well advised to do so quickly. Already they are a rarity in this arid place and their numbers will only decline as global warming dries out their home further. And – as if frogs don't have enough to deal with – where conditions warm and dry and rainfall patterns change, their breeding cycles and general survival will be turned upside-down.

Six hotspots have been identified by the *Frog Atlas* as regions where high numbers of threatened species exist. Three of these are in places where the pages of my life have been written. The Amatola Mountains: the place of my youth. The Western

Cape lowlands: the scene for my adulthood. The KwaZulu-Natal Midlands: where it all started. My first childhood steps were taken within range of Bilbo's chirruping roadside song. Only back then, in the early seventies, no-one even knew that this frog existed. Sadly, Bilbo's family line may well not continue much further into the future.

Curlew sandpiper, one of the long-haul travellers which summers at Langebaan Lagoon
Photograph © Mark Anderson

# Missives from the Russian tundra

*Der spring has sprung, der grass has riz, I wonder where dem boidies is.*
Anonymous[1]

Every year in March, when the long, hot days of summer are just about to give in to the inevitable demands of autumn, the sandpiper and plover families start packing for their annual trip north. Like their neighbours the whimbrels, turnstones and knots, the brothers godwit and all the stilt kids[2], the residents of Langebaan Lagoon plan for a journey they have been making for longer than quill-scribbled histories can recall. They probably don't even remember why they do it. Like another species whose Yuletide pilgrimage takes them to the hot sands of Durban and Cape Town, this trip is so much a part of tradition that there is simply no other way of doing things.

When the curlew sandpiper clan receive their cue to make ready for departure, the birds head out into the mudflats left soggy and crawling with food by the daily retreat of the tide. In their tweed-coat feathers and dun-stockinged legs, they stab at the mud like hyperactive jackhammers, unearthing a pantry-full of morsels: worms, snails, crabs, shrimp, insects. They need to be well stocked up for where they are going.

Why this ritual still dictates their lives remains a mystery. It probably has something to do with the benefits of living in perpetually fine weather where the food remains abundant. The advantages of chasing down the planet's summers must – to start with at least – have outweighed the drawbacks of exhaustion and the other hazards of making the toughest long-distance migrations of any animal on the planet. Whatever the reason, the paths followed by avian generations have etched themselves into the genes of these creatures.

A swallow that summers in the Cape could probably find everything it needs in the tropics of Africa but, once it leaves Britain, it is compelled to travel almost double the distance further south[3]. There's simply no defying that cellular directive. But before you assume that birds just follow these maps by rote, think again. A Welsh Manx shearwater was deliberately 'lost' to see if it could find its home again. It was released in the United States and was 'back in its burrow on Skokholm Island off the Pembrokeshire coast one day before a letter announcing its release'[4].

Whatever the reason for making the journey, the curlew sandpiper and its many water-wading relatives of the lagoon start hoarding for the long road ahead. As the days shorten and the Sun angles towards the horizon, the change in light flicks a switch in the sandpiper's brain and it starts to gorge. This is the marvel of evolution – the signal does not come with the onset of winter or when the birds' food source has dwindled, it comes when there is plenty of time to prepare. If the cue came later when the weather dipped, there would be no chance of stocking up the pantry in time for the great trek. Like a binary switch clicking to 'go', a physical change in the hypothalamus (where hormones and involuntary actions are governed) leaves the animal insatiably hungry. It eats, and eats, and eats. The same trigger alters the chemistry in its blood so that the calories go straight to the waistline. It packs on the weight until nearly 50 per cent of its body mass is fat. Give it a couple of weeks and the bird is fuelled up and raring to go.

You would have to be made of strong stuff to tackle a trip like this: the curlew, all 90 little grams of it, 40 grams of which are fat reserves[5], will fly 14 000 km to get the best of those long boreal summer days. In six weeks it will track the west coast of Africa, cross the equator, leapfrog around the bulge, skip across the Strait of Gibraltar, pass over fjords where Vikings once sailed, and follow the northern European coastline[6] beyond the seventieth parallel, clocking up seven days of total flying time. It's destination: the treeless plains of the Taimyr peninsula – the Arctic tundra.

Birds have mastered the art of long-haul travel. Since their ancestors first jury-rigged reptilian scales into lightweight instruments of flight, they have defied gravity to cross the planet's most formidable oceans and landmarks. They are an aerodynamic *tour de force*, which makes a mockery of our feeble attempts at propelling great, lumbering chunks of metal into the sky.

The Arctic tern makes a 32 000 km[7] round trip every year, moving between its breeding grounds in the Arctic and its austral summer stomping grounds in the Antarctic. That's close to a global circumnavigation every year. If a six-seater Cessna were to make a similar trip, it would burn over 5 000 litres of fuel and need to touch down 26 times (every 1 200 km) to fill up[8]. The Blackpoll warbler (*Dendroica striata)* is so fuel efficient that, if it were an aircraft, it would get the equivalent of over 300 000 km to the litre during its 90-hour trans-ocean marathon dash for the south.

A human performing at a similar metabolic rate would have to run the four-minute mile for eighty continuous hours[9].

Wandering albatrosses are born with itchy feet and are engineered for sailing the westerly winds of the Southern Ocean. A youngster will spend its first seven years on the wing, only coming in to roost on its home island when it's ready to mate. Masterfully crafted by evolution, its wings lock into position, meaning its heart rate is probably lower when it is gliding on the wind than when it 'rests' on water where it must paddle[10]. Bar-tailed godwits appear to fly a staggering 13 000 km from Alaska to New Zealand in one mighty leap across the Pacific without a single stop[11].

In mountaineering-speak, 8 000 m above sea level is the start of the death zone. Even sucking on a canister of oxygen, the human body slumps into depression. It becomes physically exhausted, can't catch its breath, feels dizzy and faint, experiences headaches that feel as if the skull will crack open at the temples, and seems unable to heal injuries. This is no place to harbour an infection. The summit of Mount Everest, the highest point on Earth, reaches up to 8 850 m – that's intruding into the path of high tropospheric jet streams. Even weather patterns find the towering Himalayas impassable. The mid-summer average up there is minus 19°C.

Bar-headed geese (*Anser indicus*) do not see the problem. Every year they pop across the Himalayas on their way from their breeding grounds in Tibet to India and back again. They are regularly seen flying high above the summit of Everest. These are the supremos of high-altitude migration. And, to date, the highest ever bird strike to an airplane happened at 11.5 km up, when a Rüppell's griffon vulture was sucked into a jet engine above the Ivory Coast of West Africa in 1975[12].

You would think that with this phenomenal amount of mobility, birds would be less likely to succumb to the rigours of a changing climate. Plants cannot uproot and move – if they want to shift their range in response to climate change, they die out on the inhospitable side and slowly sow their seeds on the conformable side, allowing subsequent generations to carry on the legacy as they track climate and physically shift their range. Terrestrial animals, unless caught up against a geographical barrier, can move more easily, particularly if they have corridors of favourable habitat. Birds, surely, could hop on the next connecting flight and find somewhere else to live[13]. In theory, that's true. But birds have elbowed their way towards the apex of the food chain, so any changes further down the ladder will manifest quickly in their communities. That's why they are a good indicator of change.

When the curlew sandpiper arrives at its breeding grounds in the high Arctic latitudes, it has the benefit of long hours of daylight, few competitors and not that many predators. But, that close to the pole, it only has a tiny window of opportunity. From the moment it arrives in early June, the gap closes fast. Timing is everything. Adult curlews plan their arrival to coincide with the first snow melt, where a felt-like covering of mosses, lichens, sedges, grass and some herbs so typical of the coastal tundra peek through for the first time in nine months. The ground is frozen solid but a slow thaw increases the gap between the nesting ground and the permafrost. Three months later, when the time comes for the stragglers to leave by late August, there will be about 45 cm of sludgy soil insulating them against the permanently frozen earth beneath[14]. The curlews have little more than two months[15] in which to peg out their nesting place, pair up, mate, brood the egg, get the chick on its own two feet, feed themselves up and leave for the south. Before long, the hazards of a polar autumn will swoop down with frigid, inky nights and inundations of snow that cover the place for two-thirds of the year[16].

How they manage to use this gap best depends on various things: how fit they are upon arrival – if they have been slowed by bad weather or had poor refuelling stops along the way, they will be exhausted; and how well they fit their timing to the cues of nature. Coastal waders from Langebaan Lagoon that have chosen to breed along the shores of Medusa Bay in Taimyr, about 1 900 km from the pole, have to time their nesting and chick hatching with a tight but slightly elastic relationship between the weather conditions, snow melt and the flurry of insects, mosquitoes and other invertebrates that the downy newborns must survive on for the subsequent weeks.

One thing is critical – in the peak of summer, which is by no means the drawn-out affair we are used to in South Africa – the tundra has exactly one week in which temperatures climb and bring with them a flush of invertebrates. The parent waders need to be certain that they are settled in and brooding their chicks to hatch in time for the rush because the chicks' healthy growth is directly linked to the availability of food[17].

These high Arctic breeders produce tough little offspring – curlews are on their own from the moment they hatch, leaving the nest and foraging as soon as they have removed the egg muck from their eyes. There's no sitting back with gaping mouths, bossing their parents about. Thanks to yoke reserves stored in their belly, they have enough energy at shell-break to make it through those first tortured moments of life. Day-old curlews have been found a kilometre from their nests[18], foraging between stubborn patches of snow. From the moment they breathe their first, they are fattening themselves up for their migration south. If the chicks are lucky, they will be two months old when they start their journey, many are even younger[19]. There is no place for apron strings this close to the pole.

Timing is critical for migratory birds: for millennia they have responded to cues from nature, linking their breeding to the peak time of their food sources, whose habits in turn are shaped by cues of their own. Those birds which receive at least part of their migration prompt from ambient temperature are already responding by migrating earlier in a warmer world. While some birds do not appear to have changed their migration time at all, others definitely have. The Eurasian curlew (*Numernius arquata*) advanced its arrival time in Deeside, north-eastern Scotland, by 25 days between 1974 and 1999. Little ringed plovers (*Charadrius dubius*) and whimbrels (*Numenius phaeopus*) have shifted their ETA by 'six to twenty two days per decade and three to six days per degree Celsius in relation to the mean January to March temperatures'[20]. The ringed plover (*Charadrius hiaticula*) appears to be laying its eggs earlier.

In theory, earlier arrival for high latitude breeders could be beneficial because it gives them a bigger window of opportunity[21]. Increasing temperature could also mean more food on the insect front. But, true to form, nature is never that simple. Even if the birds arrive early, there is no guarantee that the snow will have cleared. Ironically, a warmer world means increased precipitation in places. In these parts, this falls in the form of snow, which will take longer to melt and may stall nest-building. There is also no guarantee that their food will be waiting for them when they need it. Insects, a major food source for many birds, have a short, sharp lifespan and may respond quickly to different cues. Some hurry through many generations in one season, meaning they may be able to adapt to change faster than species with longer life cycles. Already some insect distributions have shifted with climate change[22].

Scientists refer to this as a discoupling of species interactions[23]. Depending on the players, it could run through entire food chains: the bird needs the insect, which needs the plant, which needs the rain. In a system where one species is triggered by day length, another by temperature, and a third by seasonal rainfall, climate change could juggle these cues and it is likely that the whole system will become discombobulated.

While the links of the food chain are snapping, the very tundra where breeding takes place may be shrinking. Arctic tundra is hemmed in towards the north by sea and in the south by a ring of boreal forest. As these latitudes warm, boreal forests will shift northwards, colonising the tundra. Hemmed in by sea, this habitat will steadily shrink. Worse still, as sea levels rise, the tide will push inland, flooding coastal breeding sites, altering salinity and eroding coasts. A 65 per cent decrease is expected in the Arctic tundra[24]. In Alaska, the average temperature has already been showing an increase of 0.5°C per decade for the past 30 years, during which time shrubby plant growth – once uncommon to the area – has begun to take root in places.

High Arctic breeders can expect an increase in warming within the next one hundred years that is 40 per cent above the global average[25]. As the tundra becomes milder, predators from the more temperate regions will also migrate north. Already the Arctic fox has given over much of its southern turf to the red fox in Canada[26]. Other predators such as sparrowhawks, badgers, hedgehogs, weasels and polecats are expected to move into high Arctic breeding grounds. And, of course, so will humans. The frigid north has kept away all but those adapted to its conditions. As these become more favourable, you can be sure that a wave of settlers will move into this new frontier.

The Avian Demography Unit (ADU) at the University of Cape Town has been counting the birds of Langebaan Lagoon for 30 years. When it projects its figures onto a graph, the general trend in bird numbers is downward in spite of the occasional flux between seasons. The curlew sandpiper, knot and sanderling show the greatest declines, but the grey plover and turnstone are not far behind. For some reason, these conditions seem to be favouring the larger-bodied whimbrels and bar-tailed godwits whose numbers have increased over this time[27]. It just goes to show how unpredictable species response can be to changing climate. Doug Harebottle at the ADU says that, while this decline has not been linked conclusively to global climate change, it is fair to say there is a strong correlation. As with so much else, the evidence points towards the likely culprit, or at least one that is a partner in crime with many of the other factors threatening birds in a world where increasing numbers of humans live. Climate change is the single greatest threat to the planet's wading birds[28] and the days of the great northern wildernesses are numbered.

Until such a time, however, the curlews of Langebaan Lagoon will continue their annual pilgrimage north, flouting all political boundaries carved into maps with ink, blood and nationalist zeal. They will skip over border posts and bypass capitals with neither visa nor passport. They will still confound us with the wonder of their migration. When August starts to close up shop in the Arctic, the new generation of curlews will be fattened up and ready to fly south all on their own. There will be no adults left to guide them south and to mentor them in the ways of trans-continental flight. They will tune into an instinct that has the blueprint of the night skies fixed in their heads, their senses tuned to magnetic fields, and an inherited memory of landmarks and the setting sun.

Forget bird-brained. These winged creatures will reach Langebaan Lagoon a few weeks later a whole lot more worldly than your average infant. The youngsters will arrive to find their mothers just recovering from their own trip home, having left the chicks as soon as they could stand on their own. But their fathers will have been home for a lot longer. You see, the males only stick around in the Arctic for about two weeks, just long enough to get the eggs settled into the nest before they fly back south. Whichever way you look at it, a 28 000 km round-trip is a long way to fly for a shag.

# Chapter 7

A rare bloom of 'Darling Ivory' orchids in the renosterveld scrub
Photograph © Anton Pauw

# Castaways in
# a dry ocean

*Earth laughs in flowers.*

Ralph Waldo Emerson, writer and philosopher, 1803–1882

There's truth in the old adage about never ignoring the fine print. If Charles Darwin had done so, he would have missed his chance to make it into the annals of scientific history. He must have been pedantic about detail because, when he described the finches of the Galapagos Islands, offshore of Ecuador, he read nature's intricate messages scribbled in the curve of their beaks and varied plumage. Darwin noticed that communities of finches across the islands shared enough common physical traits and behaviours to suggest they were related, but they were evidently different enough to be set apart from one another as individual species.

Darwin's theory, as he formulated it, was promptly shelved for two decades. You see Darwin had trained to become a clergyman before his portentous trip on the *Beagle*, which delivered him to the Galapagos finches, so he was well versed in the religious Victorian sensibilities that his theory was bound to offend. It was only with the arrival of the talented young Alfred Wallace on the scene in the mid-1850s, whose own papers in a natural history publication warned he was speedily reaching the same conclusions as Darwin, that the erstwhile want-to-be preacher finally sent his book – *On the Origin of Species by Means of Natural Selection* – to print. So it was, as the older ram squared up to the new rival in defence of his territory – itself a Darwinian process, a theory was unveiled which would forever change our perception of the human condition and all the existential crises that go with it: the evolution of species through a process of natural selection.

What defines the work by both Darwin and Wallace is that their study ground was islands. By virtue of their isolation, island ecosystems are unique in the most weird and wonderful ways. The principles of island life gave Australia its spring-loaded marsupials, Papua New Guinea its bandicoots, the Arctic its great white bears, Indonesia its Sumatran rhino and komodo dragons, and Japan the stocky, dun-and-black spotted kitty that lives exclusively on Iriomotejima Island. Separated from the abundance of continental life for long enough, these creatures are free to amble down the strangest evolutionary paths.

If islands were to choose a mascot, it would have to be the flightless bird. Without the inspiration of terrestrial predators to propel them skyward, some islands' avian inhabitants grounded themselves permanently, relegating their wings to ornamentation. Consider Mauritius's dodo; the Galapagos Islands' flightless cormorant; New Zealand's kiwi and the kakapo parrot, which now lives only on three islands offshore of the Land of the Long White Cloud. Then there are those that were simply too big to defy gravity: New Zealand's several moa species, and the elephant birds of Madagascar that ranged from 'ostrich-sized up to the largest of the *Aepyornis* species, which stood ten feet tall, weighed half a ton, and laid eggs the size of soccer balls'[1].

Sadly Darwin's astuteness and the marvel of island biogeography were not around in time to save the dodo, which had already been extinct for a century and a half by the time the landmark *HMS Beagle* expedition headed for South America in 1832. In fact, the last recorded dodo was seen in 1681, having succumbed to the appetites of hungry sailors and marauding animals that they brought with them to the island.

New Zealand's moa were hunted into the grave by Polynesian immigrants and their domesticated animals after their arrival in the late thirteenth century[2], and had mostly given up the ghost by the time Captain James Cook visited the area in 1769. Bony remnants hint of twenty species – all extinct – which varied between one and four metres in height. Madagascar's biggest elephant bird is remembered as the legend of the *rukh* in *The Thousand and One Nights*[3], but it was wiped out by Indonesian immigrants. By the time Arab traders encountered Malagasy residents on the island in the ninth century, only stories of the giant birds remained.

Many of Hawaii's birds were still using their wings while their landlubber cousins elsewhere were being eaten into the afterlife, but that didn't help much. The Pacific island had a veritable aviary of 125 to 145 birds that occurred nowhere else in the world. Some were flightless (one resembling the dodo, another like the New Zealand moa), while others soared through forests more diverse than we could imagine without having seen them firsthand. Today, biologist Edward Wilson records in *The Future of Life*:

> Almost all of these endemic forms are now extinct. Only thirty five of the original species of birds still exist, of which twenty four are endangered, with a dozen so rare they may be beyond recovery.

While the island's flightless birds were being feasted upon by Polynesian colonists, its vegetation was stripped for agriculture. Pigs, rats, feral cats, goats, cattle, snails and ants brought to this region by humans then ate their way through anything that stayed still long enough. Exotic plants snaked their way over their more fragile indigenous counterparts. Wilson explains:

> Humanity, when wiping out biodiversity, eats its way down the food chain. First to go among animal species are the big, the slow, and the tasty. As a rule around the world, wherever people entered a virgin environment, most of the megafauna soon vanished. Also doomed were a substantial fraction of the most easily captured ground birds and tortoises. Smaller and swifter species were able to hang on in diminished numbers.[4]

Madagascar, Hawaii, Mauritius, New Zealand and many more areas are museums to the 'trail of *Homo sapiens*, serial killer of the biosphere' which 'reaches to the farthest corners of the world'[5].

If you are ever out shopping for an island, you will find two kinds on offer: the leftover shards of continents torn apart, such as Madagascar and the Falklands; or those formed when geological actions pushed sea floor or volcanic matter up above the ocean surface. A couple of basic rules apply: continental islands are generally closer to the mainland and start out with all the necessary basics to continue evolution (plants, animals, and so forth). Ocean islands, however, are often vastly more remote from other occupied land masses and start with nothing whatsoever. To paraphrase the words of writer David Quammen: as far as species evolution and extinctions are concerned, continental islands start with everything and have everything to lose; ocean islands start with nothing and have everything to gain[6].

Madagascar began shearing away from Africa about 240 million years ago, but it clung to the rest of Gondwanaland's India, Australia and South America for another 150 million years until it finally split away about 90 million years ago to become the fourth largest island on the planet. Like Noah's ark it took with it a host of species, mostly dinosaurs, originating from the mainland but destined to become extinct with the rest of their kind 65 million years ago. The theory is that most modern vertebrates to inhabit the island must have migrated there after Gondwanaland fractured. Small animals might have clung to '"rafts" of vegetation', while others may have crossed the Mozambique Channel during times when sea levels dropped, travelling 'by land and sea along a chain of submerged highlands northwest of the island'[7].

Many animals have strong links to species on the African continent, which suggests a possible origin for the immigrants. But 'elephants, cats, antelope, zebras, monkeys and many other modern African mammals apparently never reached Madagascar', while many rodents, lemurs, carnivores and hedgehog-like tenrecs must have made the crossing[8]. After millions of generations and years of reproduction in isolation, these creatures changed into a magnificent collection – 80 per cent of the plants and animals here occur nowhere else in the world[9]. This includes the 40 species of lemurs, from the tiny lesser grey mouse lemur, weighing in at only 60 g, to the bow-legged, ring-tailed lemur. The ruffed lemur looks like some stoic Victorian jester while the slimline sifaka darts through the forest canopy in impressive 10m jumps. There were many more now-extinct species of larger lemur – including the giant arboreal lemur, which may have been something like the modern-day sloth. 'No fewer than fifteen species of lemur, one-third of the total fauna, disappeared when the human tide washed over Madagascar', states Wilson[10].

The plants or animals that find themselves on islands grow and mutate in isolation. For this reason, islands work a fantastical magic on the species they hold to their bosom. Wilson and his contemporary Robert MacArthur described the dynamic equilibrium of an island ecosystem in the late 1960s. They argued that an island would reach a dynamic equilibrium where its species diversity represented a balance between new species arriving and older inhabitants becoming extinct. As long as the factors driving arrival and departure stayed the same, the ecosystem should remain in equilibrium[11].

Distant islands have fewer species while closer islands have more species because of their higher likelihood of successfully crossing the deserts of salty water. The theory also argues that larger islands will have more species and lower extinction rates than smaller islands.

Either way, all types of islands have simpler ecosystems than the mainland and, by extension, fewer mechanisms to counteract catastrophe. Think of it as having a weaker immune system, purely because these islands have not been exposed to as wide a variety of diseases as they would be if they were connected to the mainland. On a continent a species might also be able to pack up and move more easily if the neighbourhood became too rough. On an island it's not that simple especially if, like the dodo, an animal has discarded air travel for the luxury of picking its livelihood from the forest floor.

Technically speaking – in ecological terms, anyway – expanses of salty water are not the only geological features that can cause a piece of land to be isolated into an ecological island. Mountain peaks, rivers, differences in climate or soil can do much the same thing. This is where another kind of island is introduced – the landlocked one.

* * * * * * *

Leo Tolstoy remarked once what a strange illusion it is to assume that beauty is goodness. Tramping through renosterveld brings exactly this to mind, for it is the ugly step-sister to fynbos. But renosterveld is by no means any less special and it is probably even more in need of conservation. This dull grey-green vegetation is dotted through the Cape like an archipelago of islands and one day its story will ring with the same tragedy as that of the dodos.

Renosterveld acquired its name from the innocuous little shrub 'renosterbos' (literally 'rhinoceros bush', or *Elytropappus rhinocerotis*), which was probably so named because of the black rhino that once grazed where it occurred. The earliest settlers in the Cape – the pastoralist Khoikhoi and later Europeans – would have seen large game on the foothills of the Cape Fold Mountains. Herds of bontebok, eland, zebra-like quagga and bluebuck once grazed these slopes along with rhino. Sightings of such large herbivores should have been the first clue to the ill-fated future of renosterveld because the big game fed where the nutrients were high. And, where nutrients were high, cattle herders could graze their domestic stock and settlers could grow their grain, fruit and canola crops.

The nutritional superiority of renosterveld areas over fynbos, where no grazers much larger than the size of an average domestic dog were found, results from the more mineral-rich nature of the parent rock that makes up the granite and shale soils which, over the millennia, have been exposed by geological processes. This richness, along with the nutrient-trapping clay nature of the soils and lower rainfall, combined to form the perfect habitat for what is not quite fynbos and not at all succulent karoo. 'Where rainfall is greater than about 600 millimetres, renosterveld is replaced by fynbos, and where it is less than 250 to 300 millimetres succulent karoo takes over'[12]. Renosterveld is a kind of no man's land between the two that looks like nothing but is loaded with life.

Dr Anton Pauw, a specialist in the ecology and evolution of plant-animal interactions, points out that the true glory of renosterveld lies hidden beneath the shrubs. Here, concealed from view for most of their lives, is a latent kaleidoscope of flowering bulbs. To botanists this represents what Lagavulin does to single malt drinkers. Or, as Pauw describes, it's the connoisseur's garden. Lillies, irises, orchids and amaryllis bulbs by the hundreds are buried in the dirt, just waiting for a fire to rip through the renosterbos to give them a small window of light and nutrients to burst forth into flower. Flitting between them, in a flurry of activity, are the myriad little pollinating bees, flies, beetles and wasps whose job it is to ensure that the bulbs always have offspring.

Two things typify the bulbs of the renosterveld – the highly localised home ranges of many species, and the custom-designed nature of their pollinators. One of these unusual interactions is the green-ivory flowering orchid called *Pterygodium cruciferum* and its own special bee. Because so few of these rare plants have been domesticated, most do not have a common name. So, for the purposes of this story, and to get around that tricky *Ptery*-what's-its name, I'm going to call it the Darling Ivory.

The Darling Ivory is unique in that it is one of a few highly select species of flowering plants that does not secrete nectar to attract its pollinator – instead, it secretes oil. The concept is fascinating but the process is not well understood. What kind of physiological differences must a plant have for its flower to produce a short-chain fatty acid instead of the usual sugar water associated with insect-alluring flowers? No one seems to know.

This is a flower that evolution has chosen to leave unnoticeable. The vivid colour or pungent-scented nectar that attracts the everyday pollinator are missing from the Darling Ivory. Honey bees and flies buzz straight past it. Instead it has evolved side by side with a peculiar oil-collecting bee over millions of years. The bee is nothing like the highly social honey bee whose workers sacrifice themselves for the greater good and have a very sweet . . . um, mandible. The Darling Ivory's pollinator (*Rediviva peringueyi*) is a solitary creature, lives in clay-type ground and feeds its young on oil 'bread'– which sounds rather provincial but is one of the many fascinating interactions to be found in the renosterveld community. Its specially adapted leg hairs draw up the clove-scented oily 'reward' of the Darling Ivory flower, which the bee then mixes with pollen to make a nutritious bread for its youngsters. The bee flits from one flower to another, collecting for its larder, all the while taking with it the packages of pollen that guarantee the Darling Ivory another generation of life.

There is some infidelity in the community, though. This particular bee pollinates about 15 different oil-producing orchids and a few snapdragons. Each flower must ensure that its parcel of pollen does not get delivered to the wrong species while the bee does its rounds. Masterfully adapted, each flower delivers its pollen to a different part of the bee's body from where it is later collected by the next flower of the same species.

To be honest, this pollinating mutualism between the Darling Ivory and *R. peringueyi* is more the *presumed* relationship than one that has actually been witnessed and quantified. Pauw explains that no one has ever actually *seen* the Darling Ivory being pollinated, but this particular bee is the only one that fits the criteria for the job. This, you see, is the big problem with renosterveld – no-one knows very much about it, and not enough of the habitat remains in order to find out. Pollination by oil-collecting bees was only discovered in 1984. The bees themselves were only described in 2001. So many other questions about the Darling Ivory and its partner bee remain unanswered: for instance, how many flowers must be visited, and over what distance, for the healthy dispersal of the plant's genes to be ensured; what flight distance is the bee capable of, or where does it go during the non-flowering years when it isn't seen anywhere? These are all critically important questions because the Darling Ivory now only exists in three remaining patches of renosterveld: one near the town of Darling on the West Coast, another at Elim near Sir Lowrie's Pass, and another site that Pauw prefers not to name because of its vulnerability to poaching.

Renosterveld was once common on the west and south coast lowlands of the Cape until *Homo sapiens* arrived with their guns, plough shears, wheat, fruit and hungry cattle. First the megafauna disappeared under a volley of bullets. Extinction came fast for the bluebuck, quagga and the Cape lion that fed on them. Then, as farmers spread across the valley floor, they created an ocean of rippling grain which now laps at the feet of the Western Cape's mountains. Peeking out of this beige sea are a few islands of renosterveld and whatever fauna has managed to scramble to safety. The jetsam of an evolutionary marvel, discarded and slowly dying. After 300 years, an estimated three to four per cent of West Coast renosterveld remains – most of it in little patches atop distant hills and in shadowy dales, saved from the tractor by virtue of steep slopes, rocky outcrops or marshes. Conservationists are hurrying to protect these renosterveld remnants but, in line with the principle of small islands being most vulnerable, they are most at risk of environmental decay.

Consider, by way of illustration, Van Gogh's *Starry Night*. Imagine for a second that you could take a pair of dressmaker's scissors to it and cut out each of the pulsing orbs in its oily night sky. Cut along the outline of the trees and then trim each little rooftop into neat blocks. Now, have you turned the priceless painting into an equally valuable puzzle of many parts, or do you have a pile of ruined canvas and an extremely annoyed art community? How useful is a fistful of hilltops when the species on each are isolated from one another, and when pollinating insects cannot travel from one patch to the next to keep the gene pool fresh? At what stage does an island ecosystem become so small that it becomes obsolete? The theory of biological diversity on islands holds that smaller islands support fewer species. Each time a new row of wheat laps higher up the shore of a renosterveld patch, there's less space left for survival. Think of it as the islands of Tuvalu disappearing under a rising sea level.

The prognosis for the Darling Ivory, even in the medium term, is not good. As soon as a piece of vegetation becomes fragmented, the pollinators become scarce. Pauw estimates that the smallest viable size of renosterveld is probably about 450 hectares – the size of Signal Hill, above Cape Town's city bowl. Darling Ivory's bee is still present near the West Coast site, but scientists like Pauw will probably never be able to verify the pollination mutualism between the two species because patches of renosterveld are becoming smaller and more compromised by agriculture.

Comparing remnant patches of renosterveld in city areas such at Tygerberg with others in the pastures of Darling confirms that – in spite of the huge tracts of pastures grown for cattle – there is still more biological diversity in the pastoral areas than in the cities where the 'matrix' is quite different and possibly less forgiving. Pastures, even though they have eroded vast expanses of renosterveld, are porous and allow many renosterveld species to thrive in them. But even this is changing. Monocrop agriculture is moving in with its intensive single-variety tracts and deadly pesticides. Wheat and vineyards are far less forgiving of biodiversity than pastures – and sadly the Darling area is now bursting with wheat fields and expanding vineyards as wine journals hail the exceptional 'terroir' of these rolling lowlands. The matrix is being converted to increasingly intensive forms of agriculture.

Two small reserves – one near Darling, another near Malmesbury – each have their own species of bulb that occur nowhere else on the planet. Each reserve is only five hectares in size. But the Malmesbury reserve is surrounded by wheat fields where the fertilizer run-off is insidiously changing the reserve's natural processes. How much change is taking place, though, no one knows. After all, there's no historic and scientific benchmark as so little is known about these areas.

The prognosis for renosterveld is already dire. The list of species waiting in extinction's departure lounge reads like the prophet of doom: the kukamakranka (the Khoi name for the *Gethyllis afra*); the endemic bulbs of Darling and Malmesbury; a type of buchu now found only on a single ridge near Dassiefontein; the jasmine heath with its exquisite pink flowers; the last few plants of the daisy-yellow *Moraea elsiae*, which flowers in the centre of Kenilworth Racecourse in the southern suburbs; many of the Disa orchids . . . the list goes on and on. Now imagine what happens when the stresses of climate change are superimposed upon this already faltering system? Not only will these last remaining islands of renosterveld warm and dry, but their bioclimatic envelope will shift to the south and east, pushing up against existing farms and cities.

The subject of meat is causing me to age prematurely, so conflicted am I by the ethics of the messy issue. Everything about my grinding molars and tearing incisors tells me that I am the descendant of a long line of omnivores, whose diet must include grains, vegetables, fruit and *meat*. I'm also partial to the occasional lamb shwarma or roast chicken, alongside all the other more politically correct side orders of low-glycaemic, high-fibre fruits of the earth. And while I'm no fashionista, I know that leather just feels a whole lot better than tacky plastic imitations. But no matter how neatly deboned, deskinned and 'de-animalised' our chicken breasts are, sterile cellophane packaging *cannot* make a euphemism out of the real problem – that the way in which these products reach my local supermarket often involves a less than agreeable journey to demise for the animal that finally becomes my lemon-and-herb seasoned dinner.

Yet here I sit at my computer, a slice of marmalade toast slowly cooling at my elbow, and my question is this: how many animals and plants have lost their homes and lives so that I could have my breakfast? The wheat basket of South Africa is any place where renosterveld once grew (barring a few summer rainfall wheat adaptations upcountry). Every square metre of wheat that is shoe-horned between the coast and the Cape Fold Mountains is another small patch of renosterveld lost.

If you are going to open the ethical Pandora's box of meat and leather, my dear vegetarian friends, then what do you have to say about the environmental damage wrought by your own need for nuts, vegetables, fruit, pasta, soya, ricecakes and bread? For no matter how animal-free are the products that you eat, somehow, somewhere on this globe, species have met with an untimely end to provide the monocropping and large-scale agriculture that keeps you in your daily recommended allowance of vitamins, minerals, carbohydrates, fats and protein.

But before I brand anyone a hypocrite, let me confess to my own complicity. I may only eat a few servings of meat a week (most of it bird, less of it mammal – as if that somehow makes it better), but the rest of my diet is loaded up with vegetables from the West Coast, fruit from the East Coast, tea from Kenya and chocolate from the bulge of Africa. My diet is far from wheat-free and for years I have made money out of writing about (while consuming plenty of) the Cape's best and worst wine. I am as embroiled with renosterveld's two big killers as the guiltiest party. Suddenly my toast doesn't taste quite so good and I have to wonder if there truly is a guilt-free mouthful of food left on this planet.

# Chapter 8

One of South Africa's tent tortoises, a chelonian treasure
Photograph © Marius Burger

# Sex in the slow lane

*Nature is slow, but sure; she works no faster than need be; she is the tortoise that wins the race by her perseverance.*
Henry David Thoreau, naturalist and author, 1817–1862

If an equity-minded tortoise were ever to sit down with the ghost of Walt Disney, it would probably demand reparation. In Disney's world a tortoise can lift off its shell, like a gentleman tipping his bowler to a lady. Or, if the animal tripped and toppled over, its shell might go spinning off down the pathway, leaving the owner blushing and naked. Disney hasn't given the tortoise a bad rap, but the anthropomorphised creatures that have strolled, skipped and stuttered their merry way across movie screens for decades have misrepresented their true characters to generations of impressionable youngsters.

A tortoise can no more lift its shell from its back than you could snap off your fingernails.

And if the tortoise had a beef with another stereotype, it might want to have a word with the late Grecian wordsmith, Aesop, who spun the famous old yarn about a tortoise, a hare and a great race. It's a universal tale about a contest between unequal competitors. And the slow chap wins. While the hare may be typecast as highly strung and a tad ditzy, our tortoise has filtered into the modern psyche as being slow of body and possibly a bit slow of mind.

Occupied as modern storytelling has been with perpetuating these stereotypes, it hasn't noticed that the tortoise has been quietly applying itself to one of nature's more challenging geometry problems: how to engage in a meaningful romantic liaison when its flat, bony belly must meet the domed shell of its partner. Without the use of opposable thumbs or a working knowledge of Pythagoras, the tortoise has overcome this odd feat of engineering – so successfully, in fact, that evolution has not chosen to rearrange its shape much in the 200 million years since its pre-dinosaur origins[1].

This particular journey starts in the Karoo – that scrubby, shimmering, treeless land on either side of the N1 en route to Cape Town, which many of us whiz though in air-conditioned bubbles, pausing only for a pit stop or a quick burger. The Karoo gets its name from the Khoikhoi word for 'barren' or 'dry' and it's a bleak countryside indeed. Here, across the Groot Swartberge to the north of Oudtshoorn, lies the drowsy hamlet of Prince Albert. It's a quaint little place that's chock-a-block with retired academics, artists and eccentrics and it has an abundance of fossils (the palaeontological kind).

Not far from Prince Albert is a square of land on a private farm that has been fenced off from the ubiquitous nibbling of sheep that make this area legendary. It has two small prefabricated buildings and a doorless long-drop with a great view of the mountains. Here at the Tierberg Karoo Research Centre, amidst the loose clumps of gnarled and stunted shrubs breaking otherwise bare ground, Dr Thomas Leuteritz came to unravel the secret sex life of a sand-dwelling nomad of the Great Karoo.

A long time ago – before he discovered a world where ancient, scaly creatures carried their homes on their backs like nature's own AutoVillas – Leuteritz was told that he would never amount to much. He was at the bottom of his class and struggling to make it through his exams. One of his teachers (a kindly Mr Walter Zawadski) had the foresight to suggest he be given a little extra time to complete his exam papers. By the time I met him, the German-Canadian had disproved the misconceptions of his youth, mastered his dyslexia and was approaching the tail-end of a year's post-doctoral research. To the residents of Prince Albert, he had simply become known as Dr *Skilpad* (the Afrikaans word for 'tortoise'). The creature that Dr *Skilpad* had come here to study was the tent tortoise (or *knoppiesdopskilpad, Psammobates tentorius tentorius*), close cousin to the endangered geometric tortoise (*Psammobates geometricus*).

Southern Africa is the Mecca of the chelonian world. If tortoises were currency, we would be sitting on Wall Street and the United States (US) would be a third-rate economy. There are only 43 tortoise species in the world. Fourteen occur across southern Africa, and many of these do not occur anywhere else in the world. We have more species than any other continent. Even the United States and Mexico only have four tortoises.

The problem is that virtually nothing is known about our tortoises. The US has hundreds of researchers picking apart the lives of their four species but most of southern Africa's chelonian troopers have soldiered along under the radar of science. Leuteritz and his colleagues from the University of the Western Cape's (UWC) Chelonian Biodiversity and Conservation Programme intend bringing this to an end. They plan to document the lives and times of each of southern Africa's tortoises. The need is more pressing now than ever because some could find themselves perilously close to the abyss of extinction as habitat destruction and climate change continue to bear down on them.

In October 2002 Leuteritz started the search for twenty tent tortoises – twelve females and eight males. These creatures, accessorised with radio transmitters glued to their shells, would become his only company out on this windswept, lonely Karoo plain for weeks at a time. Every three weeks he would visit the site and track down his charges. Their movements were documented to see how they used the environment and what they had eaten. Mostly, though, Leuteritz wanted to understand their reproductive cycle.

Almost a year later, once the broad strokes of their movements were established, it was time to fill in the finer details. Bobbins of thread were attached to the shells of six female tortoises. As the animals foraged, slept, sunned themselves and occasionally engaged with one another, they left string in their wake like Theseus in the Minotaur's labyrinth. Leuteritz and his assistant documented these trails in minute detail. Every so often the thread's story would be interrupted. Here and there it became wound in tight circles around a shrub or dead branch. This was usually punctuated by track marks like a rain dance padded out in the dirt. Signs of mating – or at least an attempt to do so – were tapped out in the Karoo dust.

It seems the male tent tortoises try to get it on at any chance they can manage.

'They're horny little bastards,' chuckles Leuteritz, standing over a bunch of knotted thread.

The string-trailing experiments were done in August and showed that, in spite of the late winter chill, the tortoises were quite frisky – in fact virtually every time a male encountered a female, he tried his luck at mating. The string-trailing method is useful but the best way to check whether the male has any success is to find evidence of eggs . . . as if finding the tortoises isn't hard enough. You would think that spotting tortoises in such sparse groundcover would be easy – but it's virtually impossible to find them once their parchment-and-black shells have melted into the dappled shadows, so the chance of catching them in the act of nesting is stacked towards the improbable. It's much easier just to catch the female while she's still carrying the evidence.

In the old days a female animal would be dissected to see how many eggs she had. If the goal is conservation of a species, which it usually is these days, dissection is not the most constructive or politically correct way of doing your research. For the past thirty-odd years, scientists have turned to x-ray technology to lend a hand. Every three weeks Leuteritz would collect his twelve females, x-ray the animals on a portable machine, and then visit the local hospital to have the negatives developed.

'You can see the outline of the tortoise and its bones. Because the egg shell is hard, like bone, it shows up white on the x-ray too,' he explains, pointing to the flat white disks silhouetted on the blackened transparency.

'With the sequential taking of x-rays, you can get an idea of how many times an animal has laid eggs. You can get an idea of when they have eggs, how many eggs they have, the size of eggs and, to some extent, how long they hold those eggs.'

Unfertilised 'pre-eggs' or follicles, which have yolk development but no shell, don't show up on x-rays so another form of medical technology comes to the rescue: ultrasound. The animal is suspended in a tub of water and an ultrasound probe, usually the type used by gynaecologists, is pressed against its hip or the base of its neck where its limbs emerge from the shell. Little pearls appear on the screen of a portable monitor – the next generation of tent tortoise. Then back the animal goes into the wild, all bits intact, off to lay another day.

Already Leuteritz has made valuable discoveries through his work on the reserve. There was a perception that tent tortoises didn't move during winter. People failed to see them so thought that they probably dug themselves into the ground. By examining their behaviour closely, Leuteritz has found that the animals actually stay above ground and are quite active in spite of the weather.

The male seems to move more than the female, which is part of his reproductive strategy. His smaller body size is adapted to greater movement, which allows him to cover wider areas and find more females along the way. In some species, such as the bulky leopard tortoise, the male dominates the female in size and can manhandle his way into sexual liaison. When the male is as small as the tent tortoise, he must use speed, agility and mind-numbing patience to wear down his mate. Tent tortoise females are about 13 cm in length, while the males are about 10 cm. It would appear that the male reproductive technique is to nag his way into the sack.

'Interestingly, we've found that the tent tortoise and its cousin, the geometric tortoise – although closely related – use different mating strategies. The geometric tortoise produces eggs between September and December. To our surprise, we find much longer egg production with the tent tortoise – their season lasts from October to June,' Leuteritz comments.

This strategy does not account for the higher abundance of the tent tortoise over its endangered cousin.

The tent tortoise's strategy is designed for dealing with its unpredictable, arid habitat. By producing eggs over a longer period, hatchlings come out at several different times throughout the year so, at some point, at least one clutch is going to experience favourable conditions when there is plenty of food about for them.

Tent tortoises are rugged little beasts that can survive on the smell of a proverbial oil rag. The few edible plants that they can eat in this predominantly poisonous flora are scarce. It also helps that the tortoise can last up to two months without water. And, when the water comes – unpredictable, erratic thunderstorms bringing sudden, short deluges of rain – they harvest it wherever they can. They might submerge their heads into a puddle of water and drink in great drafts. If there's too little water for that, they have been known to snort the precious liquid up through their nostrils like a cocaine user.

Leuteritz has also seen a tent tortoise use its shell in the way that humans use gutters. It will stand with its rear legs lifted higher up and its front legs pulled in. As it rains, the water runs down the shell towards the head, then drips off the shell and onto its legs from where it drinks the water. The tent tortoise's bladder is also adapted to water storage. Unlike the mammalian bladder that does not allow the body to re-absorb water, the tortoise can absorb water from its bladder and reuse it.

Like a perfect mobile home, they trundle along with their worlds on their backs, water reserves topped up, and always on the look-out for food. The ladies also carry with them all the ingredients for procreation – a larder-full of unfertilised eggs and as much semen as they can gather on the way.

It's unconfirmed in the South African species, but some tortoises and turtles are able to store semen for later use for up to a year. A male and female mate whenever they can, but in the event that they don't see their way right to enjoy later conjugal visits, the female can still produce a few clutches throughout the laying season by storing semen. The egg simply leaves the ovum, moves into the oviduct where an eager little deposit of sperm will be waiting to fertilise it. Tortoises, it seems, are ahead of the game with the world's first drive-through sperm bank. The female can then hold onto the fertile eggs for a few months (although it varies between species) before she decides conditions are just right to lay them.

Having adapted to such hardship, tent tortoises don't survive abundance. Scientists have found that this species does not do well in captivity and no one knows why. Home-fed lettuce simply does not cut it (although I probably could have told you that). This is important detail to note because if anyone were to start a breeding colony for conservation purposes, they need to know that something exists in the tent tortoise's habitat that allows the animal to survive, which is not readily available in captivity.

About 300 km to the west, as the crow flies, is the site of another tortoise research project. Here, on a private conservancy outside of Cape Town, Leuteritz's colleague Dr Brian Henen is studying the last few remaining generations of the endangered geometric tortoise.

For enigmatic reasons of its own, the geometric tortoise seems to live only in renosterveld, that dull-looking, scrubby vegetation that forms a strange no-man's land between the succulent karoo and fynbos in the western parts of the country (see Chapter Seven). The tortoise has not been found in any other plant communities and even seems to be fussy about certain types of renosterveld. Because its habitat has been carved up and sterilised by monoculture, communities of geometrics have retreated to the hills. As the vegetation is shaved away slowly by the plough shear, patches of renosterveld have become smaller and smaller and so have the communities of tortoises living there. Fragmented populations are scattered across the region from Wellington to the Ceres Valley. They are estimated to have a population density of about one or two tortoises per hectare, so a ten hectare patch of renosterveld may only have ten tortoises. Count the few remaining mounds of renosterveld next time you are travelling through the wheat fields of the Swartland and the Overberg, and then do the maths.

Henen reckons there are not more than three or four populations of geometrics remaining that are large enough to be genetically viable. A viable population is one that can persist for a long period of time – not a year, or five years, but a few thousand years. A population like this may experience fluctuations over the years but will essentially be stable over time. Henen's research site is probably the largest population of geometric tortoises left, but he is reluctant to make its location public because of the risk of poaching. This community has the minimum criteria for viability at about 2 000 to 3 000 tortoises and it simply cannot afford to lose any more animals. Any smaller community is not going to survive in the long run.

Most other populations are in the order of twenty to thirty hectares – probably meaning twenty to thirty tortoises. Scientists have a name for the precarious corner in which these communities find themselves: extinct in the wild, once a community or individual can no longer contribute meaningfully to the species' overall gene pool. By virtue of its isolation, its genes are redundant. It may as well be dead. (The same applies to species kept in private collections and held in captivity – even if the owners have permits, they are not really contributing to the conservation of the animal if it cannot breed.)

These minute hilltop communities are living museums of species existing in a genetic cul-de-sac. Before long, in-breeding will weaken some and pare down their numbers. It will be more and more difficult to recover from any natural dips in their abundance through drought or fire or flood. Ultimately, it's only a matter of time. Alarmingly, some developers have been known to use this excuse to motivate a new development. Why bother saving a species, they argue, when it's going to be extinct in twenty years from now? It's a bit like pulling the plug on a coma patient when you are not sure whether he or she is past the point of no return.

The life story of a geometric is one only recently unveiled. When a hatchling ventures from its shell, it is a bantamweight – weighing in at only 15 grams (a medium-sized chicken egg weights about 50 grams, according to my kitchen scale). For the following two to three years of life – when a human child would be learning to walk and mastering the rudiments of language – it will battle the odds with a shell no more rigid than a weak fingernail. It will use its markings as camouflage while it waits out the time until its body can scavenge enough calcium from the environment to harden its shell, which is an externalised extension to its ribcage.

The early years are precarious and only a few animals make it through their youth. By the time the human child would be settled into Grade 4, or even witnessing the distant approach of adolescence, the geometric tortoise might just be classified as an adult, aged eight to ten years. It will have another forty years (perhaps a few more) before it needs to consider retirement.

It's vital to piece together these tortoises' lives and how they slot into the overall system. If anything changes, scientists need to know what has changed, and by how much. With the geometric tortoise, it's too late to get authentic baseline information because the animal's habitat and numbers have shrunk so dramatically that all science can do is record its current status, precipitously close to being gone forever.

But, with the tent tortoise, it's still possible to glean some kind of an idea. For instance, the tent tortoise's population appears to have settled into a stasis of about one male for every female. What would happen, though, if temperature increases predicted for the area do occur? Let's gloss over the issues of changing rainfall, the impact of food supply and shifting habitat for the moment, and just look at temperature increases. Most tortoises and turtles, like many reptiles, are dependent on temperature for their sexual 'orientation'. It's called 'temperature dependent sex determination' or TSD. In mammals, sex is predetermined by genetics. In some species of turtles and tortoises, and in most crocodilians, the outcome is decided by the temperature at which the eggs incubate. If eggs incubate at a cooler temperature, the hatchlings tend to have a bias towards being male. If they incubate at a warmer temperature, more females emerge[2]. In the case of the tent tortoise, it's just theoretical because no one has studied which way TSD swings for this particular species. Theoretically, a winter batch might be more inclined to produce males while a summer batch might produce females, or the other way around, depending on how this species works.

The other thing to consider is how hot and cool nests sites are; how shady or sunny they are; or the effect of depth in the ground on the eggs. Eggs closer to the surface may be a little warmer than those deeper down. All these things come into play in the gradients or differences of temperature in eggs and the resultant sex of those eggs. If more females hatched in a warmer climate and there weren't enough males, would the population's equilibrium be thrown out of whack? Even if there were more females – meaning greater breeding stock, no one knows what might happen if the overall population shifted to, say, three females to every one male.

'I don't know if the males will be able to keep up with that,' Leuteritz says with a grin.

Tent tortoise males are twenty per cent smaller than females. They have to wait until the females are ready to mate and dash from one female to another to catch any window of opportunity. If, all of a sudden, you don't have enough males, this could throw the population into a crisis. One can only speculate how a male-dominated community would respond.

An ancient relative of the tortoise is seeing the effects of temperature switches. Nile crocodiles in the Greater St Lucia Wetland Park are the southernmost population of this toothy reptile. Their breeding ground is under siege by a weed that is foreign to these parts, the notoriously invasive Caribbean creeper, the triffid weed (*Chromolaena odorata*)[3]. The crocodiles at St Lucia tend to lay their eggs in open, sunny, sandy areas but in at least one breeding place, the laying site is heavily shaded by weed.

A study in the late 1990s into the changing conditions of the Mpane River, which feeds into the southern end of Lake St Lucia, has found that shading caused by *Chromolaena* decreased the temperature of the sand at a depth of 25 cm by as much as 5°C to 6°C compared with that of sunny areas at a similar depth. This was expected to produce a bias towards female hatchlings in the crocodiles. The decreased incubation temperature might also compromise the health and number of young crocs to hatch successfully.

Crocodiles have a limited incubation range of between 28°C to 34°C – within that range is the ideal temperature that will produce one male to every female in the hatchlings. Amongst St Lucia's Nile crocodiles, the lower end of this 'pivotal' temperature is 31.7°C. As incubation temperatures drop to below that, the sex ratio will lean towards the production of females. Below 27°C, healthy incubation is less likely – the baby crocs probably won't hatch. Through two breeding seasons during the study period, soil temperatures in shaded areas reached 26.1°C and 25.6°C. Researchers also noticed that as much as 40 per cent of breeding sites were inundated with the weed in just four years of study. They speculated that females were trying to breed in shaded areas contrary to their instinct because of a lack of suitably sunny areas. St Lucia is one of only three breeding sites for the Nile crocodile in South Africa and, if the trend is not reversed, it could become locally extinct from St Lucia.

At least the tent tortoise has the luxury of a stable and abundant population to indulge in theory and speculation. The future of its more threatened cousins, such as the geometric tortoise and the southern speckled *padloper*, is less certain. Leuteritz, using the more scientific crystal ball that computer modelling provides, has seen a bleak future for these animals. He took the eco-climate modelling produced by the South African National Biodiversity Institute (SANBI) and superimposed onto it the extent of the remaining renosterveld and the home range of the geometric tortoise and its threatened cousin, the southern speckled *padloper* (*Homopus signatus cafer*). The results showed that as the habitat of renosterveld shifted south-eastwards, it clashed dramatically with current land use practices. By 2050 it looks as if there will be nowhere left for renosterveld or the geometric tortoise to live. So if you want to see the walking dead, take a look at a picture of a geometric tortoise.

The people behind the private conservancy are already thinking into the future. They have no choice. If climate change predictions do materialise, or come faster than expected, they need the ability to maintain a captive breeding programme for the geometric tortoise. Conservationists cannot wait until it's too late before they have the know-how and technology to be able to maintain colonies outside their natural habitat. If the tortoise's natural habitat disappears through climate change, conservationists must have a plan ready. As Henen points out: 'We've never been faced with an issue like this before. What happens if a species is on the edge of extinction and its habitat is gone? You may be able to breed some of the plants. But the thing is it's not just the geometric, it's the whole ecosystem. There's a great deal of interplay between the plants and the animals. Tortoises need plants for resources and refuge; plants need tortoises to spread seeds.'

Studying animals like these requires a dedication far removed from the office-bound lives of many city folk today. It's pretty grim work at times – labouring through the blazing sunshine, with gnats and flies worrying every inch of your sweating face; or stalking across the Karoo with bitter winter gales ripping through the valleys in search of a tortoise whose radio transmitter has malfunctioned. Week after week Leuteritz, Henen and their research assistants count, weigh, record and probe the secrets of these animals before the rarer of them are gone forever. Henen's search for tortoises has taken him to desert places where he has been exposed to dangerous levels of UV – already he has received treatment for skin cancer. Leuteritz has walked an often lonely journey from Madagascar's dry, spiny forest to the Karoo's backwaters with unfailing good humour. These scientists are a rare breed, much like the creatures they study.

One evening I watched Henen searching for unfertilised eggs in one of his tortoises. The kitchen of a local guest house near the reserve had been turned into a makeshift laboratory: the kitchen table cleared of the remnants from our dinner, a plastic tub filled with water, and next to it a portable ultrasound unit rigged up to its probe. In his hand, Henen grasped another tortoise and suspended it in the water. The animal seemed strangely unconcerned – its miniature elephantine legs flailed in the water in slow motion, almost absentmindedly reaching out for a firm hold on something. When its head occasionally dipped under the water for a few seconds, its little button eyes didn't blink. A thin string of bubbles trailed up from one nostril. When it resurfaced, there was no gasping for air. There was no panic or thrashing about, none of the wild-eyed terror of a mammal being held down in water.

I was slumped back in an uncomfortable chair, willing my leaden eyes to stay open. By now it was approaching midnight after a long day of stalking through the veld, following the *ping-ping-ping* of the tortoises' transmitters. But Henen seemed to be in a kind of Zen state. All the while he pressed the ultrasound probe against the tortoise where its legs emerged from the shell, searching for chilonean treasure. And suddenly, there they were. Three little orbs of life-waiting-to-happen, floating like ghostly pearls on the screen of the ultrasound monitor.

Throughout the evening we had talked about the meaning of it, the struggle to study these animals and the personal sacrifices often involved. Now, seeing these shimmering orbs on the screen, perhaps seeing their future, Henen said quietly to nobody in particular:

'Sometimes I wonder why I bother.'

The next day the tortoise went back to the veld, with her parcels of expectation tucked safely away inside her shell. Hopefully, when they make it into the world, they will have someone watching over them as closely as Henen is now. That's if they hatch at all.

# Chapter 9

Creeping orange monsters of the deep, the West Coast rock lobster
Photograph © Geoff Spiby

# The Jabberwocky's son

*The dragon-green, the luminous, the dark, the serpent-haunted sea.*
James Elroy Flecker, english poet, 1884–1915

Twas brillig, and the slithy toves
Did gyre and gimble in the wabe;
All mimsy were the borogoves,
And the mome raths outgrabe.

Beware the Jabberwocky's son,
the jaws that bite, the claws that snatch,
Beware, fine urchin balls, and shun
the frumious abalone catch!

Yet he spread his armoured tail afan
As tiny manxome slugs he caught
So rested he by the algal tree,
And swam awhile and thought.

And, as in uffish thought he stood,
Jabberwock's son with pincers splayed,
Came worming through the seaweed wood,
And clattered as he came!

One, two! One, two! And through and through
The armoured tail went flipper-flap!
Orange-plated shell stood in good stead
And he went galumphing back.

Hast thou seen the Jabberwocky's son?
That flailing claw, that squeamsome boy!
O frabjous day! Callooh! Callay!
He clattered on his way.[1]

Peeking from beneath a brooding overhang of rock on the sea's rough and ready floor are two orange twigs. Inert and harmless they first appear. Until one twitches. The other taps. These are by no means inanimate – they are attached to a knobbly, creeping, spiky, orange, armour-plated scavenger of the deep and they are looking for a wayward urchin to happen by. The West Coast rock lobster is a suit of armour brought to life – it's all jointed legs with creeping, twiggy feelers reaching, pincers snaring, and beady eyes gleaming through the murky deep. It's a real world Jabberwock and a horror of the depths for any abalone snail that chance happens to bring this way. And it's just another one of the many improbable life forms that evolution has humoured itself with in the other-worldliness of the sea.

This particularly monstrous-looking creature only lives along the West Coast of southern Africa, along the southern Namibian coast and down to the eastern side of False Bay. At the bottom of the sea, its close to the top of the hierarchy, feeding on the urchins, snails and sea stars that graze their way across these strangely waving submarine pastures. Theirs is a world not unlike that of the blowing grasses, bleating herbivores and chasing meat-eaters of the savanna. But it is a great deal more watery.

Water. After the air we breathe, it's pretty much everything in our day-to-day survival. As you are right now, wherever you are, about 62 per cent of you comprises a remarkable coalition of two hydrogen molecules and one oxygen. No, you're not a big slushy water balloon. Rather, you're a collection of neatly packaged cells that keep this transparent life liquid neatly distributed throughout your system. So, next time you feel the need for some self-analysis, just stare deeply into a glass of water – you'll be sure to find some meaning.

Water. It's the key factor that separates our planet from the others sharing this solar system because our planet is the only one where $H_2O$ dominates the surface and shape-shifts between gas, liquid and solid. It covers two-thirds of Earth's surface. If we were able to roll out the planet's crust into one, neat constant layer – literally smooth out the deepest ocean trench to the same level as Everest's peak – the entire globe would be lost beneath a 2.5 km-deep layer of the stuff[2].

Water. Only three per cent of all this quantity of water is fresh – and, of that, most is locked up in glaciers and ice sheets (nearly four-fifths in total); another one-fifth is hidden away in underground aquifers; and the *fraction* remaining (0.3 per cent of all fresh water) is above ground and available to you and me for our eight recommended daily glasses of it[3]. The rest of the planet's water makes up the most mysterious and inaccessible places on Earth – so difficult are its high-pressure depths for us to explore that they might as well be on another planet. This is the constantly swirling, thrashing, splashing, pounding, churning, sloshing, dripping, spraying, misting, relentless body of mineral-rich water in which life first evolved: the ocean.

The ocean is a vital mechanism in the great planetary thermostat, taking the Sun's heat from the air above it and redistributing it around the planet through its constantly circulating currents, gyres, swells, eddies and tides. Imagine a globe – better still, if you have an actual globe somewhere, go and dig it out now. If not, raid the fridge for a roundish piece of fruit or grab your kid's soccer ball and imagine the continents and oceans on it. Pay close attention to the Atlantic, Pacific and

Indian Oceans as their expansive girths straddle the equator. You may want to draw some of these features onto your fruit (best not on the soccer ball, though).

Remember what was discussed earlier about the planetary winds – fired up by the principles of convection – that churn up the surface of the ocean into an ever-restless state. Now, give your globe/ball/piece of fruit a spin so that it turns from left to right (that's from west to east, so the Sun appears to rise in the east and set in the west). Recall how we discussed that the Earth's spin deflects these surface currents, to the left in the Southern Hemisphere and to the right in the Northern Hemisphere. There is then a 'drag' effect on the layer of water immediately below the surface, and the next layer down, and so on. This 'Ekman transport' effect produces a gyre or gigantic spiral of water in each hemisphere of each ocean.

Now go back to your imagined or actual globe (you may stop the spinning now) and picture how these circular gyres spin clockwise in the north and anti-clockwise in the south of each ocean. Both carry heat away from the equator towards their western edges and towards the colder, higher latitudes close to the poles. Once there, they pick up a chill which they carry back down the eastern edges of their oceans. Looking at South America, southern Africa and Australia, each continent's western coastline is touched by the cold eastern reaches of the gyres that spin past them.

The Benguela Current is one of the four major eastern boundary currents in the world. Prevailing winds, associated with the South Atlantic High Pressure System and the eastward-moving cyclones at these latitudes, combine during the spring and summer months to produce a relentless south-easterly wind that blows off the continent and out to sea. This process sweeps the surface water of the ocean, warmed by the Sun, away from the coastline, causing cold waters to rush up from the ocean floor to take its place. The ocean floor, like any self-respecting compost heap, is loaded with nutrients: everything from the carcasses of whales to fish faeces and last season's seaweed float down once their time has come and decompose on the seabed. When the washing machine action of the wind stirs the cold, bottom-level water up to the surface, it takes this thick, highly nutritious sludge with it, turning the water murky and dark. Closer to the surface and bathed in light, marine algae or phytoplankton – the microscopic 'grasses' of the ocean's food chain – bloom spectacularly in response to this organic fertilizer. Enter the zooplankton to graze down this new growth of green. Fish and whales then feast on the zooplankton, while marine mammals and birds feed on the fish. And so the food chain bursts at the seams.

This process of cold-water upwelling has created the ocean's greatest fisheries: on the Grand Banks off eastern Canada, the Georges Banks off New England, along the coast of Peru, the English West Coast, the south-western United States and, of course, off the West Coast of southern Africa. The actual number and variety of species on the West Coast may not be as great as on the east, but the sheer volume of fish, measured in biomass, far exceeds the rest of the warmer east. For this reason, the fishing trawlers work these western shores. Tropical waters, which are warm and relatively nutrient poor, are watery deserts by comparison with these chilly upwelling regions[4].

Like cows thriving in deep pastures, life in these waters depends on the blooming of algae. And the blooming of algae depends on the upwelling. The upwelling depends on the wind. It's one great interconnected roundabout that turns like clockwork.

Meanwhile, far below and entirely oblivious to the oceanic machinations above its head, the rock lobster sets about the business of lurking in caves and snatching up unsuspecting sea creatures. Its life starts out as one of about 200 000 eggs – of which only one per cent will make it to adulthood – that are glued beneath a membrane on the underside of its mother's tail. After about nine months it will hatch from the egg and be cast adrift on the ocean. There it will spend three or more years bobbing about in the upper surface of the water, where it starts out feeding on plankton and eventually jellyfish. It will moult several times during this period, changing shape slightly each time from transparent bug to armoured adolescent.

As the rock lobster approaches its adult form, it will begin to swim up and down between the surface and the ocean floor, scouting about for a future home. Once it settles into adulthood, the female will first begin to mate at about seven years of age. This is a slow-growing creature that will shrug off its spiny orange carapace once a year, swell out its flesh by absorbing water, wait for its larger shell to set, and then settle back into its new casing. It grows little more than one to five millimetres per year, if even that. So it continues until the ripe old age of 40, if it is lucky enough to evade the fishermen's traps or the hungry maws of the octopus, dog sharks and seals.

But we know nature is highly unpredictable. At times, the gentle ticking of the system in cadence is overwhelmed by nature's occasional and dramatic extremes. When the system turns on the rock lobster, it happens in the most spectacular and grizzly fashion. In April 1997, as if in some kind of cultish suicide pact, the rock lobsters of Eland's Bay, just north of Saldanha, abandoned their watery home and marched up onto the beaches as one. There lay 1 700 tons of lobster stranded on the sand, their tail-flapping and creepy-crawling becoming feebler by the hour as the sun rose above them and baked them in their shells. In less than a day of leaving the sea, they turned black, fetid and highly toxic.

As it so happens, the blame in this case lies squarely with the offshore winds. After a full season of harassing the coast and drawing up the cold water from the bottom of the ocean like mercury to a magnet, they began to falter towards the close of summer. It happens like this, where a pulse in the wind proves catastrophic. The water is loaded up with nutrients; the phytoplankton grow as if they are on speed; but a pause in the south-easter means they are not blown out to sea. The algae multiply as if it were their last day alive, turning the water sludgy with excess bloom. Soon the tide turns rusty red, bleeding toxins into the water that may poison some creatures and even prove lethal to people.

If the across-shore currents conspire to sweep this notorious poisoned red tide closer to shore, the Jabberwocks are doomed. As the plants run their course and deplete the available nutrients in the water, the algae die and sink down through the water to the ocean floor. Here bacteria begin their work on breaking down the algae to enable it to continue on its way through the nutrient cycle. But this decaying process gobbles up the available oxygen in the water. Without the usual wind action to churn up the sea and keep it oxygenated all the way down through the water column, life on the ocean floor begins to suffocate.

Lobsters, breathing through their gills, are extremely sensitive to low-oxygen events. As this oxygen-deficient water is swept in from deeper waters, they move shoreward to escape it – up into the shallows of the inter-tidal zone. The closer the low oxygen water intrudes into these bays and shallows, the more it draws the lobsters towards their impending demise. If the event is mild, it will only stress the creatures captured in the low oxygen pockets, inhibiting their growth rate and stifling reproduction. But a bad event will drive the lobsters shoreward. The Jabberwocks will follow the tide as far as it will take them, but once the water retreats, the creatures are left stranded between suffocation by water or suffocation by air. It's not only lobsters caught up by this swirling, invisible death: swimming crabs, cuttlefish, squid and octopus, mussels, sea snails and deep-water fish all succumb to loss of oxygen[5].

Five mass walkouts occurred during the 1990s, bringing 2 263 tons of lobster out of the sea. Three of these events were the worst ever recorded in South Africa[6]. In the most extreme cases, once all the oxygen has been depleted, the bacteria are left to work in an anaerobic environment. The consequence is that they start churning out hydrogen sulphide, which bubbles its poisonous gas up through the water, tainting it once more. The result can be catastrophic. Entire bays have been sterilised by this acid – seaweed, shellfish, fish, lobsters; you name it, they die. All that remains is the noxious smell of rotten eggs and a sterile bay. Life here will eventually recover, but in its own measured time.

Lobster walkouts and red tide events have been on the increase in recent years. Various other things have also been changing. The deep waters of the Southern Ocean are warming – by as much as 0.17°C between the 1950s and the 1980s, and the sea surface temperatures around southern Africa have climbed by about 0.25°C per decade for the past 40 years or 1°C since World War Two[7]. Dr Barry Clark, one of the few marine specialists in South Africa to have pondered the fate of marine life in a changing climate, comments that while the global circulation models predict an increase in air temperature of 1°C to 3°C over the region, there is less confidence when it comes to predicting changes in sea surface temperature. It is understood, though, that this will lag behind air temperature by about 20 years.

It seems, too, that an increase in the intensity of upwelling is occurring in many of the world's 'upwelling centres', including along South Africa's West Coast. Clark says that currently the most accepted explanation for this event suggests that night-time cooling is being stalled by the increase in greenhouse gases in the atmosphere. This causes daytime heating of the land to increase, leading to a greater difference in the pressure gradients between the sea and the land. This gears up the offshore winds a notch or two, which speeds up the upwelling process.

Besides the unpleasantness that an increase in wind means for West Coasters who loathe the pesky Cape Doctor, the predictions are that a continuation of this accelerated upwelling will bring greater amounts of nutrients to the surface, which means the phytoplankton receive the equivalent of a fuel injection. If the increased upwelling is followed by late season pulses in the winds – as is anticipated – then low oxygen events can be expected to increase. This means more lobster walkouts, more death of life down below, and a dearth of some of your favourite deep-water fish for dinner. Increasing frequency and intensity of red tide blooms and low oxygen events along our West Coast are a sure thing.

But the low oxygen events are only expected to occur when the wind slacks off. Surely, for the rest of the season, the increased productivity of the algal bloom will mean greater food for the system – more zooplankton, more fish? Surely this is the answer to the demise of fisheries globally? Not likely. Like so many things in nature, near-surface fish such as anchovies and sardines need just the right conditions in order for their youngsters to thrive. Of course, if the offshore wind is too weak, there won't be sufficient upwelling to ensure a plentiful supply of food and the fish larvae will struggle to survive. Weaker air movement also means that the water will not be churned well enough to bring the food and larvae into contact with one another.

Too strong a wind, however, can be equally damning to the next generation of fish as it will blow the spawn far out to sea and away from their food source. Higher winds will also churn up the water too much, breaking up the pockets of food on which the fish larvae must feed and dispersing them widely. Or they will mix up the water so much that the phytoplankton will be driven too deep where, deprived of sunlight, they will not be able to photosynthesise. The larvae also find it difficult to catch their prey if their world is churning around them like soapy water in a washing machine. So, like every good story of nature, the wind must be just strong enough but not too strong. For fisheries off our West Coast, that's estimated to be about 9 to 11 knots (or 18 to 21 kilometres per hour).

Clark says that wind stress is projected to increase over southern Africa, with the consequent greater possibility of harmful algal blooms. Combined with changes in other phenomena, this 'may further complicate an already unclear picture'. Wind stress is certainly expected to disrupt the spawning process of near-surface fish in the area. While some larger species such as sardines may be better equipped by virtue of their size to migrate to more suitable areas, smaller fish such as anchovies may be better able to adapt to environmental change by remaining in the same area[8]. Either way, it's yet another environmental change whose outcome we have little control over.

And, if you think you wilt in heat waves, you can be grateful you are not a fish. When it comes to changes in temperature, numerous marine species are among the most vulnerable. Mammals have evolved a clever mechanism whereby they can regulate their internal temperature to a level that is relatively constant and comfortable, making them a little less dependent on their environment to keep their core body temperature stable. Cold-blooded animals are not quite that efficient. They are dependent on the temperature of their surrounds to keep their metabolism in full working order. Locked into a specific set of thermal requirements, they respond immediately to change.

Like many terrestrial species, fish are expected to respond to changes in the temperature of the environment around them by shifting their distribution. Fish from the tropics are likely to push south along our coast. Clark says that many fish from temperate latitudes might try to head for deeper waters, but could find their migration inhibited by water currents. Our bristly orange lobster has already started pushing eastwards, around the False Bay coast and into the breeding grounds of the abalone snail. There the lobster is gobbling its way through young abalone even before the poachers can get to them[9].

Mercifully, lobsters will be spared the ravages of rising sea levels, which will bring increased erosion, tidal flooding and destructive storm events to our coastlines and estuaries, even as saltwater seeps steadily into the groundwater inland of the high tide mark. Neither will lobsters be touched by the decrease in fresh-water runoff from the continent as rainfall declines. But many other fish and estuarine and coastal species will be compromised by these changes. Clark wrote in the journal *Marine Policy* that the 'implications of climate change are likely to be at least as severe as those of uncontrolled exploitation (of fisheries), and will almost certainly become manifest within the next 10 years if they have not already begun doing so'. Thankfully a strict seasonal quota system is in place to limit the number and body size of lobsters that are taken out of the sea each year, which in a small way may help to stabilise their population numbers.

Red tide is not quite so obliging. It does not discriminate – young and old, breeding females and ancient males, every generation of lobster is driven out towards their sun-baked, sandy death. And the more that happens, the tighter the control of the commercial exploitation of these marine crawlers will have to be. Suddenly I don't care much for having lobster on the menu – somehow the idea of eating thermidor when the lobster is about the same age as me just doesn't seem right. I would rather the West Coast rock lobsters were left to age away slowly in their underworld where caves and rocky overhangs hide their prehistoric form – a place that's *brillig, where the slithy toves do gyre and gimble in the wabe.*

> But beware, Jabberwock, your son!
> Your jaws that bite, your claws that snap,
> Beware, those red-stained swells, and shun
> the frumious death tide trap.

One, two! One, two! And through and through,
Armoured tail limply flips and flaps,
That orange-plated shell's for naught,
There's no galumphing back.

I have seen the Jabberwock, his son,
The flailing claws, the squeamsome boy!
Oh travestous day! Callooh! Callay!
He died right where he lay.

Reefs: an underworld psychedelia, beset by a warming world
Photograph © Geoff Spiby

# Citadels under siege



It was not my finest moment. My lily-pale legs hung suspended in the water, swaying slightly, a bit like bait. The rest of me clung to the side of a sun-worn dhow while I coughed and hacked my way to recovery. The skipper, bless his heart, tried to afford me a margin of dignity by working on an imagined problem in the hand-cast fishing line that hung over the other side of the boat. His sole crew hand stared fixedly at the horizon.

'Ok?' asked the skipper, when it sounded as though I had found my breath.

'Um . . . mmm . . . I'll try one more time.'

I squashed the mask back over my face, bit down on the snorkel and wriggled back into the water, shaving some skin from my elbows on the way.

Breathing face down in water goes entirely against millions of years of evolutionary process. Our ancestors have not done it since those early fish relinquished the comfort of aqueous suspension and headed for dry land. It's a most primal survival instinct not to breathe in water and, even though we have now developed apparatus to defy it, it still feels wrong to do so. I've been a water baby longer than I can remember, but hopping over the side of a rickety fishing boat into the Zanzibar Channel with a snorkel shoved in my mouth was like stepping right back into nappies. After a few minutes, though, the hyperventilating gave way to deep, quavering breaths and the panic retreated to the fringe of my consciousness. It was lovely down there, but not quite what I expected. It was a bit like Picasso during his Blue Period – monochromatic and, well, very blue. Not having snorkelled before, I didn't know if that was the norm, but the sun had retreated behind the clouds and the vivid colours I expected from the tropical reef were not there.

The rigours of descriptive prose demanded more than that, so I later turned to books to wash in the missing hues. And there, in all their glory at my local library, was every full-colour coral I could ask for. If I didn't know better, I would swear that whoever made coral did it on acid. Brilliant yellow top hats with waving jester tentacles, decorated with slashes and dots of deep mauve. Crimson sea fans of the most delicate tracery. Anemones, lithe and undulating, belly dancing in the current, intoxicating to all but their attendant clownfish. Hard, branching corals like some sort of forest relic, petrified with time and then painted riotous colours. Purple sponges with fur-like trim, flirting with the water. Shapes like fossilised daisies and rosebuds. Frilly mushrooms. Spiky marshmallows. A submarine psychedelia of tubes and towers, bristles and mounds.

Coral reefs must be life's most magnificent spectacle. A cartooned medieval village dressed up for Mardi Gras, this community lives and grows just beyond the waves in the planet's warmer ocean waters. No length of yarn has been spared; not a frill, trim, sequin or bead omitted. The tragedy of tragedies is that – like the world's dripping glaciers – the reefs have become something of a cliché in the saga of global warming. The truth is that these day-glow underwater communities are under siege.

The bizarre little creatures responsible for these coral hamlets are only found in the warm waters of tropical oceans, less than 100 metres deep, where enough sunlight penetrates to stimulate plant and animal growth. Other species of coral are found in colder parts, even in the frigid waters of the poles, but many are not as sociable and they are certainly more reserved with their colour palette.

Compared to the colder, upwelling parts of oceans with their murky, nutrient-rich waters and productive fisheries, tropical waters are warm and clear. While they have many more different species than the cold west, the biomass of tropical waters is much lower. Without the rushing up of marine compost, the tropical sections of oceans are like hot, watery deserts. Corals are able to live in these nutrient-poor conditions and reefs become something of an oasis for sea creatures in these watery wildernesses. They are regular taverns of the sea where creatures can feed, rest and breed in some measure of comfort.

So much life in oceans remains a mystery because of the difficulty of marine exploration. Coral is not the least of these mysteries, but the reefs are more than just picture book material: they have been called the rainforests of the oceans. In sheer number of species, the reefs equally match the Amazonian forests – the most abundant and biologically diverse ecosystem in the world. The reefs may support as many as one million species of animals and plants, only a fraction of which have been documented and described. Conservation International reports:

> Among the best known groups are at least 5 000 species of fish, over 10 000 species of mollusk (sic) and more than 800 species of reef-building corals. Approximately 30 per cent of marine fish species occur on coral reefs.

Reefs occupy about 284 000 square kilometres, mostly along shorelines. They act as nurseries to the fish that feed millions of mouths (mostly in developing countries), provide building materials and medical source products, attract tourists, and shield coastlines from the ravages of wave action and erosion[1]. The problem with reef builders is that they are pernickety about where they live: they can only grow where there is sufficient sunlight reaching their working parts, but not too much; they prefer tepid water between about 23°C to 30°C[2]; the water that surrounds them must be at a suitable chemical equilibrium to enable them to build their limestone structures; and they simply won't tolerate a home fouled with sediment of any sort.

Reefs have been taking a hammering for years. Over-exploitation for building materials; physical damage by fishing nets, boat anchors and souvenir-collecting divers; pollution from sewage and agricultural runoff; and sediment from erosion on the mainland are all compromising the survival of these slow-growing aquatic communities. Already 25 per cent of the world's coral reefs have been destroyed or badly degraded owing to human activities and global warming, and the 'prognosis for their survival is grim without a major global conservation effort'[3].

In terms of global warming, coral's most immediate enemy is rising temperature – not so much a gradual increase over years, but frequent episodes of temperature extremes predicted to come with the general trend in rising temperatures[4]. Repeated sudden, short spikes in temperature may prove deadly after prolonged hot spells, administering lethal doses of overheating to corals. It would be something like dunking your finger repeatedly into your steaming coffee until the skin blisters.

'Gradual heating is bringing coral communities to their bleaching threshold,' says Professor Michael Schleyer of the Oceanographic Research Institute (ORI) in Durban. 'When a sufficiently hot temperature spike comes along, lasting for days, it can push them over the threshold.'

It's this that strips the colour – and often the life – out of coral as bleach removes stains from your whites. To understand bleaching, we need to look into the strange workings of coral itself. It's a bit of a mind bender if, like me, you are not biologically inclined.

Corals are animals disguised in plant fancy dress, living as one with plant material woven into their skeletal structure. A hard coral, for instance, is an animal that will start out life as a tiny larva, not unlike many other young invertebrates such as mussels or clams. It will then be cast out into the water by its parent. After a short-lived adventure at the whim of the current, it will fix itself to the surface of the reef, often the dead exoskeletons of previous generations of coral, and begin to metamorphose.

As it grows into adulthood, the animal builds a skeletal structure by extracting calcium carbonate, essentially the component of limestone, from surrounding seawater. It lays this down according to the blueprint dictated by its DNA. Fixed to the ocean floor, it then eats the tiny animals – zooplankton – floating in the ocean around it. Zooplankton supplies the animal with nutrients such as phosphorous, but not with its total daily calorie requirements. To obtain the food it needs to stay alive, coral has to rely on greens: not to eat, but rather to feed it.

This is the weird bit. During the hundreds of millions of years over which these animals have evolved, they have entered into an inseparable coalition with a type of single-celled algae called zooxanthellae (pronounced *zoo-zan-thelee*). These algae weave themselves into the skeleton of the coral and go about photosynthesising, as plants do. They take carbon dioxide diluted by the water and, along with the energy of sunlight streaming through the water, produce amino acids and carbohydrates. Most of this – up to 95 per cent[5] – they pass on to their hosts, the coral. Like any pot plant, the algae need nutrient fertilizers, which they receive in the form of ammonia and phosphate from the corals' waste products. In this way, the two – which look like one to the casual eye – keep each other well fed. So successful is this symbiosis that it places the coral and zooxanthellae in the upper echelon of primary producers in these richly diverse oases of the sea[6].

Reef-building corals provide the reef infrastructure, zooxanthellae supply the window dressing, and all the other creatures move in: snails, fish, slugs, starfish, sea cucumbers, urchins, anemones, flatworms, crabs, clams, seahorses, shrimps. Zooxanthellae are the finest artists of coral reefs – they provide the flamboyant colour. If they die or are forcibly evicted, they leave behind mute, limestone skeletons, tombs of their former hosts.

Any number of factors could upset the algae and trigger a 'bleaching' event – sunburn, disease, unbalanced salinity, or sediment increase. But the biggest culprit in mass bleaching is heat. Schleyer explains what happens inside coral when the water around it heats up:

> Anything that experiences an increase in temperature will find its metabolism increasing. In this instance, the plant cells of the algae go into overdrive during the photosynthesis process. A byproduct of the light

reaction in photosynthesis is oxygen. When the metabolism of the plant cells is stepped up, they produce an excess of free oxygen radicals. This irritates the host tissue so much that the corals then either expel the algae or digest them.

Literally overnight the corals will bleach. Because they are dependent to such a great extent on the photosynthetic products of the zooxanthellae for food, the corals will slowly starve without this algae. It may take a few days, but eventually the coral will die. Some people find a bleached reef beautiful, Schleyer says, something like a snow-covered landscape. But really it's just a graveyard. Corals may recover if the bleaching is only partial. But, if it's total and the animals die, eventually the reef will crumble into a rubble bed of deceased life.

Sea surface temperatures have increased by almost 1°C in some tropical regions over the past one hundred years[7]. However, 1979 seems to be a pivotal date for coral: prior to then, there are virtually no reports of mass bleaching events, even from well-observed research sites[8]. This could be due, in part, to the fact that people were not looking before then. But since 1979 six major episodes have occurred with 'associated mortality affecting reefs in every part of the world'[9].

Coral biologist Ove Hoegh-Guldberg, from the School of Biological Studies at Sydney University, argues that the temperature increases over the previous century have brought corals perilously closer to their upper thermal limit in certain parts of the ocean[10]. This means that slightly warmer years are now more likely to administer a lethal dose of heat, thus triggering mass bleaching.

The El Niño Southern Oscillation (ENSO) is a natural cycle that originates in the Pacific Ocean and interrupts 'normal' weather patterns around the globe. Typical conditions in the Pacific see strong westerly trade winds driving warm surface water towards the west. The winds literally slap their rain-bearing waters up against Indonesia, lifting the sea level there half a metre higher than on the other side of the ocean along the Ecuadorian coast. Meanwhile, Ecuador's coastline sees the kind of nutrient upwelling spoken of earlier.

During an El Niño event, the trade winds slack off. This interrupts the usual heat transfer, displacing it to the east instead. Besides the fact that this stalls the nutrient upwelling in Ecuador and Peru, usually collapsing their fisheries, it also means a build up of heat in the centre of the Pacific. This heat accumulation ripples throughout the planet's oceans, increasing sea surface temperatures in places and changing weather patterns. El Niño events bring flooding in Peru but drought in Indonesia, Australia and South Africa. They are regular, cyclical events that occur about every three to seven years – sources disagree somewhat on their frequency. Either way, global warming is expected to increase the frequency and intensity of El Niño events. For reef-supporting oceans, an El Niño event usually means a spike in sea temperature that can make life rather difficult for corals.

In 1998 a humdinger of an El Niño heated waters throughout the tropics, producing 'hotspots' where the sea surface temperatures exceeded the regional maximum by more than 1°C. What followed was the most severe mass-bleaching event on record[11]. Hoegh-Guldberg wrote:

> For the first time, coral reefs in every region of the world recorded severe bleaching events. In some places, such as Singapore, bleaching was recorded for the first time. Many massive corals (which may live for well over 1 000 years) have died as a result of the 1998 event, including some with an age of up to seven hundred years.[12]

This bears repeating: the year 1998 may have brought the most extreme conditions experienced by those corals in seven hundred years, possibly more.

Along the east African coast, the most vulnerable reefs lie between the latitudes of 12° to 16° South, followed by the cooler latitudes north and south of there[13]. South Africa does not host much of this extraordinary marine ecosystem – only about 50 km of coral reefs perched precariously on the fringe of habitable conditions for reef building. These reefs are the tail end of the reefs that run down the coast from equatorial east Africa and past Mozambique, trailing off along the KwaZulu-Natal north coast. The corals along this coastline draw just enough light to survive.

The pH balance is such that their process of extracting calcium carbonate from the water and depositing it in their skeletal structures is slow and laborious. Since they are at a relatively high latitude, these reefs did not reach the bleaching threshold of those closer to the equator. Thus they were spared death by bleaching during the 1998 El Niño[14].

Through the coral programme at ORI, Schleyer and his colleagues have been recording changes in the reefs of Sodwana Bay. Their findings have shown that the sea surface temperature over the past ten years has crept up steadily by 0.27°C per year, confirming a short-term and steep temperature increase in the region[15]. Since the South African corals are already at the marginal limits for their development, warming of the sea has encouraged hard coral growth at the expense of the more prevalent soft corals. But this won't last, states Schleyer. As atmospheric carbon dioxide levels increase, so the ocean's uptake of carbon dioxide by surface water will rise, increasing the carbonic acid content of the water. This will shift the water's pH balance to the wrong side of the chemical equilibrium needed by corals to lay down their calcium carbonate structures. Once the temperature increases beyond the standard that promotes coral growth in Sodwana, the building of hard corals could diminish once again. Temperatures are already approaching the coral bleaching threshold for the region[16] and, if warming continues, these reefs could follow the same fate as their northern neighbours.

The ORI researchers have already established the critical bleaching temperature for Sodwana's corals: about 28.8°C[17]. They discovered this when a bleaching event occurred in 2000 after a period of elevated temperatures and exceptionally clear water where more sunlight penetrated the water to the reefs. The reefs took a staggering blow.

What hope, then, do the next fifty years hold for coral reefs? Corals are adapted to extremely specific habitat requirements, so they have painted themselves into a bit of a corner. The rising sea levels expected with global warming might normally not have been a threat to coral. If time allowed it, the corals could simply have built their way up towards the light. But, with their building abilities impaired by other changing conditions, they may not be able to keep up with the rising ceiling above them.

Joan Kleypas from the National Center for Atmospheric Research in Boulder, Colorado, predicted in 1998 that the sea in which coral reefs grow could become disrupted in a way that will tear their world apart. By 2050, sea surface temperatures are expected to rise by 2°C – this will shift their ideal temperature conditions poleward by 18 degrees of latitude. Meanwhile, the increasing atmospheric $CO_2$ absorbed by oceans will upset the chemical equilibrium in the water and shift ideal conditions for coral building towards the equator by 15 degrees of latitude[18]. In places, coral reefs will find themselves in a tug of war as conditions pull them in opposite directions, possibly leaving their inhabitants homeless, hungry and facing a bleak future. Schleyer asserts that the computer modelling used to reach this conclusion was a bit simplistic. Further recent research indicates that the corals may be more resilient than we expect them to be, 'not an unexpected trait in nature'.

Adult corals cannot shift their location because they are anchored to the reef. If species hope to shift their geographical range, they will have to depend on their juveniles being able to drift with the current. Even so, there is only so far that they can go. In this scenario, they may not find the conditions they need to meet their physiological requirements.

Schleyer thinks a small chance exists that some species of corals and zooxanthellae are more able to withstand change than others and might survive the transition, leaving at least remnant pockets of corals. We can only hope against hope that there are some species out there with sufficient resilience.

Modelling done by Hoegh-Guldberg indicates that:

> The thermal tolerance of reef building corals are likely to be exceeded every year within the next few decades. Events as severe as the 1998 El Niño event, the worst on record, are likely to become commonplace within 20 years. Most information suggests that the capacity of corals to acclimatise has already been exceeded and that adaptation may be too slow to avert a decline in the quality of the world's reefs.[19]

If a coral animal becomes bleached, it may not necessarily be fatal. Loss of zooxanthellae may only be partial and they may return before the coral starves. Some colonies appear to be more resilient to bleaching than others and could at least keep the reef partially functioning until it can recover more fully. The more extreme the bleaching event, however, the greater the mortality of corals. And the more frequently these events happen, the more dire their future.

Hoegh-Guldberg states:

> There is little doubt that present rates of warming in tropical seas will lead to longer and more intense bleaching events. Given the behaviour of reefs over the past twenty years, most indicators point to the fact that mortality rates are likely to rise within the next few decades to levels that may approach almost complete mortality.[20]

According to the IPCC, within the next twenty years sea surface temperatures could well exceed coral's thermal threshold – 'from 2020 onwards the average bleaching event is likely to be similar to, or greater than, the 1998 event'[21]. By 2040, coral bleaching will be worse than the 1998 event almost every year[22].

Temperature not only kills coral but inhibits its reproduction and:

> Persistent bleaching events such as those predicted for twenty to forty years' time may mean that corals that are not killed will fail to reproduce – with obvious consequences.[23]

The impact that the inhibition of reproduction will have on the reef-dependent species of invertebrates and fish, and the birds and marine mammals which feed here, remains to be quantified. But the loss of this richness of life in these otherwise unproductive waters will ripple out across the food chain and be felt in the bellies of those fisher folk whose families have relied on reefs as a source of protein for hundreds of years. Around the world, 15 per cent of the global population lives within 100 km of reef ecosystems[24].The reefs yield six million tons of fish catch[25] and account for 25 per cent of all fish caught in developing countries[26]. With their function as a major source of protein for millions of subsistence families undermined, food insecurity among some of the world's poorer nations can only increase further.

• • • • • • •

Floating face down in the Indian Ocean with a plastic tube feeding me air, I hovered over a garrisoned, turreted coral citadel awash with different shades of blue. Only this one looked as though it had been under attack. Below me, washed into the gullies and channels of the reef, was a graveyard of deceased corals. Limestone skeletons of species whose ancestors pre-date the dinosaurs lay where they had been swept by currents and tides into rubble heaps on the ocean floor. Testimony, perhaps, to boat anchors, fishing nets or heat waves.

Above this macabre scene, the sun remained tucked behind the clouds and the murky blue pressed in on my fuzzy periphery. I felt suddenly aware of how alone I was out there, with no one else in the water as far as the eye could see. I felt a tingle up my spine and it was enough. Between their scratchy English and my non-existent Swahili, I pleaded with my Zanzibari fishermen companions to get me out of the water. They leaned over the side of the boat and gestured for me to place a foot in each of their braced hands. I hoisted myself from the water, like a cork out of a bottle, and landed with a thump on the boat. Again, not one of my finest exits.

'*Asante, asante sana,*' I muttered in relief. The dhow began to buck and agitate as the two men hoisted the triangular canvas sail up the mast. Swiftly I wrapped myself up in a *khikhoi* as we tacked back to the beach, as much to escape the ravages of UV as out of respect to my hosts. It was Ramadan, after all.

# Chapter 11

Forests of quiver trees are marching south
Photograph © Leonie S Joubert

# Dwarfs and giants of the dry lands

*It is not the strongest of the species that survive, nor the most intelligent,
but the one most responsive to change.*

Charles Darwin, naturalist and author, 1809–1882

I don't like deserts – hot ones that is, not the polar variety. Nothing about my freckled skin or forest-tempered sentiment is equipped for their unforgiving extremes. I feel unsafe and more than a little lost in a desert – and no one likes to be the outsider.

I was heading north on the N7, that shimmering umbilical cord that links the fairest Cape to dozens of dusty farming towns along the arid West Coast. The summer Sun ironed out the contours, leaving the landscape featureless and dull. Silver-grey mirages danced below a bleached and washed out sky; a single wisp of cirrus frayed the horizon. The mercury had long since ascended into the upper forties and the dryness leached moisture from every pore. Stuck behind a long-haul truck probably heading for Springbok or Upington, my car crawled up a mountain pass with the speedometer needle hovering below thirty. My sweat-drenched shirt clung like an insecure child. The best I could coax from the radio was static and a religious debate. This is probably as close to hell as I'm going to get this side of the River Styx, I thought, wondering how much worse it would have been to tackle this journey in an ox wagon.

Three hundred years ago many immigrants fled the saturated, over-traded economy of young Cape Town to seek out prosperity in the new frontiers of southern Africa. Landless, penniless and jobless, a new generation of *trekboer* harnessed their oxen and headed into unmapped territory. Most began the push towards the lush grasslands of the Eastern Cape, the Highveld and the East Coast. Others found themselves on the road north into the near-desert conditions of the West Coast and Northern Cape where, until then, the only humans who knew this land were darker-skinned hunter-gatherers.

As the new arrivals made their way across the plains between the coast and the Bokkeveldberg, 300 km north of the Mother City, their steel-rimmed wagon wheels ground, snarled and crunched over fields of quartzite gravel. They named this place the 'Knersvlakte' – the gnashing plains, which they first must have thought of as a desolate little hell. Water is scarce here. Scrubby vegetation seldom reaches above knee-height and the summers stretch into endless, parched months. The austral winter brings a short respite of cold-weather rain. This is the gateway to Namaqualand's legendary wild flower displays. Conditions have not changed much and the Knersvlakte remains as inhospitable now as it was for the *trekboere* three centuries ago.

But deserts are only bad news to creatures that have not evolved there. They provide a niche for any plants or animals entrepreneurial enough to adapt and make this place home.

There is one such plant community that has carved out a space for itself here on the Knersvlakte. In amongst the white quartzite stones lie a host of well-disguised, chubby little dwarf succulents. The quartzite keeps the ground cool by reflecting sunlight and provides camouflage for these plants that have taken on the appearance of their surrounds.

In summer, when the ground bakes and the rains retreat to the east, they shrivel back into the ground under a cover of gravel and their own dead skin. Here they stay, dormant, until winter's rains bounce them back into their full, voluptuous curves. In celebration they throw out showers of colourful petals. To these natives, the Knersvlakte is paradise.

After thousands of years, desert conditions have tailored this species to retain water in the same way that a camel's hump stores fat for the lean months. The succulents are efficient little operators – they siesta all summer long, like bears hibernating through winter. By shutting down their metabolism and coating their green-grey reflective leaves in a waxy sunscreen, they endure the searing desert summer. Their roots skim water from just below the surface of the ground, whether from dew or mist.

When the rains do come, the succulents burst into a flurry of growth. But just when they start blooming, the winter sun angles itself lower in the sky, sending in gentle rays at an acute angle. The plants compensated for this by evolving bulbous, cushion-shaped leaves that hug the ground and absorb the solar energy like sponges. Their woody seed pods open welcomingly with the first sprinkles of rain, allowing their seeds to be shot hither and thither by the impact of the shower – a clever device to ensure their seeds are only released when conditions are favourable for germination.

Hidden as they are from the world – as much by their own camouflage as by the climate that keeps intruders away – not many people know that the Knersvlakte harbours among the richest succulent communities in the world. At least 150 species are endemic to the area[1]. The wider succulent karoo of the West Coast is itself a masterpiece of nature, containing the largest concentration of succulents in the world, more than half of which are endemic. South Africa (both the summer and winter rainfall areas) has about 3 500 species of succulents, roughly 80 per cent of which are endemic to South Africa.[2] The majority of these are crammed into this island of arid, winter rainfall. Like the rich diversity of fynbos that grows just south of here, the succulent karoo has been identified as a 'biodiversity hotspot' – rich in diversity but most threatened by, among other things, climate change.

A little to the north-east of the Knersvlakte, over the Bokkeveldberg near

Nieuwoudtville, can be found another charismatic denizen of the dry lands. *Aloe dichotoma*, the distinctive quiver tree or tree aloe, is a giant in a sea of Lilliputian scrub. It towers over its surrounds, reaching a handsome two to three metres at full height. As a seedling, it starts life looking not unlike most young aloes – a rosette of spear-like leaves emerging from the ground. Over the years, as this crown grows taller, a stem of sheer, solid wood develops. The young bark is a beautiful, silky silver-beige wrapping over a fibrous core. Eventually the tall, slim tree splits, sending out a second branch, then a third, then a fourth, until it grows into a neatly topiaried ball of aloe-leaved growth over a constantly widening stem. By old age – which could be as much as two hundred years or more – the tree takes on a curious hour-glass shape with its ample trunk, slim midriff and neat canopy.

San people, who have migrated through these parts for thousands of years, used the tree's branches as quivers for arrows. Settlers used hollowed-out dead trees as makeshift fridges to store 'water, meat and vegetables'; the 'fibrous tissue of the trunk has a cooling effect as air passes through'[3]. Quiver trees thrive in the dry Nama and succulent karoos. Their range stretches from the Namaqualand near Nieuwoudtville and Loeriesfontein in the south to Brandberg in the Nama Desert; from Prieska in the east to Swakopmund in the west.

Tough and resilient, quiver trees store water reserves in their stems and leaves and can endure years of drought conditions. Recently, though, disturbing news emerged from Namibia suggesting that all might not be well in the *kokerboomwoude* (quiver tree forests) of the desert. Large numbers of quiver trees were dying out in the north of its range. Farmers in the area reported that in some communities of trees as much as half their numbers had died off. No one could explain why. The South African National Biodiversity Institute (SANBI) decided to compare the results of bioclimatic modelling with the state of quiver trees to see if there was any correlation between the two. Conservation biologist Wendy Foden was sent off into the desert to find the culprit behind the silent deaths of the quiver tree.

Starting in South Africa and moving north, Foden and her colleagues found communities of quiver trees across their distribution. They looked at every possible cause of mortality: fungus on the leaves, daily pressure on the land from livestock, grazing by drought-hungry animals, and baboon damage. They counted the percentage of live and dead trees in each population and then had to establish a time scale in which these fatalities had occurred. Here the trees came to their assistance. Because quiver trees show distinctive physical characteristics at different stages of life – between sapling, adolescent, adult and aging geriatric – they are their own timeline. Knowing this, Foden had a way to measure conditions over the previous few hundred years.

Mature trees flower and disperse seeds every year. However, climatic conditions are not always conducive to germination so a crop of youngsters may only appear and survive every few years when the conditions are ideal. Looking at a stand of quivers, it's possible to calculate the gap years between periods of favourable conditions for germination. In some communities they found smooth transitions, where all age groups were present. But others showed that there had been few crops of seedlings in recent years. Worse still, some communities did not have any plants under adult size.

Quiver trees start looking like adults at about 100 to 150 years of age. Foden's team did not find any youngsters, indicating that it had been a long time since quiver trees in those areas last experienced conditions where they were able to reproduce successfully.

Skeletons also help to record the history of the trees. The intensely dry conditions here mean that deceased trees may remain where they die for decades – in some cases even up to a century or more.

In the Richtersveld, which straddles the border between South Africa and Namibia, Foden and her group found 60 to 70 per cent mortality rate. Further north, in valleys near the Brandberg, a few hundred kilometres north of Windhoek, they found graveyards of trees with hardly a living aloe in sight. Not only did they notice that trees were dying, but they observed a behaviour that had not been documented before. The branches of some quiver trees came to a sudden end, in a stump, as though their hands had been severed at the wrist. This was dubbed 'auto-amputation'. Quite distinguishable from baboon damage, it seemed the trees just drop their leaves off, leaving behind a stump that is rounded at the end.

Foden's team found a strong correlation between this leaf-dropping and the mortality rate throughout the communities. Both factors followed latitudinal gradients that indicated greater die-back was occurring the further north the team went in the aloe's distribution. When the trees started shedding leaves, it meant they were on their way out. It also showed that long-term stress was the likely cause of death. Sometimes an amputated branch would send out a shoot below the stump and begin growing again – but this only happened when a rare time of abundance saved the tree from drought. It appeared that neither disease nor predation could explain the damage and death of this iconic desert aloe.

When Foden's team included the factors of altitude and latitude in the equation, a pattern emerged. Trees in the far north and at lower altitudes showed the highest mortality, while those that occurred further south and at a higher elevation were healthier. Just as environmental scientists have been predicting, these trees are showing signs of taking refuge in envelopes of comfortable habitat. They are retreating to higher altitudes, on cooler mountain slopes, while the low-lying trees are dying. The trees also appear to be 'moving' southward as their bioclimatic envelope shifts south.

Until now, the quiver tree's southern range has probably been limited near Nieuwoudtville by higher rainfall and fungus. But, as the desert moves south, so are the trees. When we talk about plants shifting their range, bear in mind that they are not like animals which can simply uproot and physically move. The only way for a plant to shift its range is to become extinct at one extreme of its range while seedlings begin to propagate successfully at the other extreme. The quiver tree appears to be dying out in its northern range in Namibia and thriving down south in areas such as Nieuwoudtville in South Africa.

In case any doubt emerged about the interpretation of the data collected by Foden and her colleagues, her team resurrected photographs taken of quiver trees during the previous century. Painstakingly they visited the site of each photograph, found exactly the same vantage point, and re-shot the picture. Comparing the old pictures with recent ones, they were able to confirm their findings. By counting trees in the photographs, they were able to see that populations in the far south, near Calvinia, showed an increase of about 110 per cent, while populations along the Orange River showed huge decreases. They found similar changes in the distribution along altitudinal gradients. This, asserts Foden, is highly convincing evidence that a changing climate is the culprit. It's the most obvious smoking gun.

Anecdotal evidence also supports the findings of Foden and her team. One Namibian farmer recalled growing up on a farm where quiver trees grew right from the farmhouse in a valley, up a nearby mountainside. Today, the first aloes to be encountered on that mountain occur about two thirds of the way up its slopes.

These findings, compared with climate records, show that there is a strong correlation between the mortality rate of trees and temperature increases dating back to the 1960s. Naturally wetter and drier periods may have occurred during this time, but the overall trend is towards drying. Temperature records from weather stations across the aloe's range in Namibia and the Northern Cape showed a mean decadal increase in temperature that is almost three times the global mean temperature increase for the twentieth century.

The conclusion of Foden's study is that the desert is moving south. The behaviour of this tree aloe confirms the theoretical projections about shifting bioclimatic envelopes. Before anyone began asking what climate change would mean to the arid west, these trees were already on the move and have been for at least thirty years, if not more. Foden comments that this is probably the last place you would expect to find climate change happening. An arid system, one might think, is more likely to handle an increase in warm and dry conditions. The quiver tree is an abundant species with a wide distribution and broad climatic tolerance. Yet, despite being so hardy, it is dying out in the hotter part of its range. More importantly, until now, this kind of plant behaviour had been limited to the realm of theory and computer modelling.

'This is a real wakeup call,' says Foden. More relevant, she points out, is what the quiver tree's experience indicates about the many other species living in these parts, particularly the succulent karoo's localised endemics. These plants do not disperse well, mostly only as far as a raindrop can propel their seeds. It's for this reason that they only occur in such small areas with highly specific climate needs.

Researchers at SANBI have had a closer look at what might happen to the succulents of the Knersvlakte in a warmer world. A series of transparent, open-topped chambers were erected around small patches of succulents during the spring and summer of 2002/2003 to simulate the hotter, drier conditions that the plants can expect to experience there in the next half century. Four months later, after exposure to an average 5.5°C hotter than the ambient temperature outside the chambers (which is close to one of the predicted increases in the area by 2080), the plants showed dramatic signs of struggle. Dwarf and succulent plants in the chambers showed a two- and five-fold greater mortality than those outside of the chambers[4].

This experiment exposed the plants to an abrupt increase in temperature, denying them time to acclimatise, which might explain some of the 'extraordinarily high mortalities'[5]. Nevertheless, many plants in the hotter parts of the chambers received lethal doses of temperature. The results from looking at the chemical properties in the leaves of those plants that survived showed that the plants' photosynthetic productivity decreased in the heat. Similar experiments on plants in forest, grassland and tundra biomes elsewhere on the planet showed that plants were stimulated to greater productivity by the heat. Not so here. These findings suggest that these rare plants are already close to the ceiling of their ideal upper temperature.

Along with Namibia's hyper-arid environment and the high levels of endemism further south, the predicted warming and drying will worsen the extremes that are naturally occurring here, making it one of the regions most vulnerable to this change. Heating up and drying out will move the winter rainfall-dominated succulent karoo biome southwards within the next 50 years, reducing its size and exerting strong pressure on the plants in this area.

The future of species that are able to move with their shifting climatic envelope looks more rosy. In an academic exercise, the potential range shift of the species in Namibia – where the quiver tree enjoys an extensive range – was modelled and the species reclassified according to the IUCN Red Data listing process. The species that were able to migrate well emerged as the least threatened by climate change. This exercise showed that 'more than half of the total number of species could at least be classified as vulnerable by 2050 and 2080.'

Of those species that were unable to migrate in the computer simulation, 25 per cent were reclassified as Critically Endangered by 2050, with a range contraction greater than 80 per cent, while five per cent were classified as Extinct. These percentages shift significantly by 2080, with 30 per cent becoming Critically Endangered and 13 per cent Extinct. This exercise shows 'the decisive role of migration in impacting on species threat status . . . (w)ith an assumption of zero migration, almost half of the 834 species modelled are classified as Extinct or Critically Endangered in Namibia by 2080, but with perfect migration assumed, only a third are Extinct or Critically Endangered, and almost half are classified as Less Threatened, even by 2080'.

But some plants will find it extremely difficult to track climate envelopes. For this reason, the Knersvlakte – along with the rest of the succulent karoo – has been identified as one of the environments on the planet most threatened by climate change. As the desert moves into these parts, the succulent karoo will be pressed up against the southern mountains and most probably will have the life squeezed out of it. SANBI bleakly states this about the region:

> Areas that support the succulent karoo today will become so arid that only the hardiest plants of that biome will be able to survive. The main part of the country where the predicted climate might support succulent karoo vegetation is the Agulhas Plain. It is highly unlikely that plants will be able to migrate from Namaqualand and the Richtersveld, across the Cape Fold Mountains to this coastal plain in the southern Cape. The succulent karoo biome faces an extremely fragile future.[6]

Months later, as I write at my desk in Cape Town, the words are spilling onto the page to the hiss of winter rain falling outside. I am 300 km south of Vanrhynsdorp and have to marvel at how mild and gentle the conditions are in Cape Town compared with the rigours of the succulent karoo. But the truth is that the desert is pressing south and I wonder what the view from my office window will be like when I turn eighty.

# Chapter 12

Royalty at 46° South, the king penguin
Photograph © Leonie S Joubert

# The frigid islands


*Not many have seen [the Prince Edward Islands], and few ever want to. They thrust their lava peaks out of that vast sea where the world's wildest weather is born. Their name is synonymous with storm, disaster and death.*

John H Marsh, *No Pathway Here*, 1948


In the 1970s scientists working on Marion Island, deep in the sub-Antarctic, would trek on foot into the island's hinterland and ski down the volcanic slopes above the snowline. They remember the white-capped cones floating on a perpetual sea of ice and snow. Today the highpoints of Marion Island remain chilly and inaccessible reaches, but now black lava and red scoria peaks have replaced the frosty landscape. The tongue of ice, the last vestige of a glacier, has retreated into the lava. A steady rise in mean temperature has slowly eroded the ice plateau. Over three decades it has steadily melted, dripped into the scoria, and refrozen inside the porous rock.

Marion's coastline is spread with deep, peaty mires. These smelly bogs were notorious for swallowing people up to their waist, armpit or even neck. Today they are drying out and becoming shallower. Now a walker who loses concentration while navigating the mires will only go in up to the knee. One of the grass-like plants, *Uncinia compacta*, which prefers drier mires, is spreading faster than ever before. A heat-sensitive plant, the rare Kerguelen cabbage, is retreating to the cooler south-facing slopes as its numbers succumb to fungus, rising temperatures and competition from other plants that are thriving in a changing system. Records going back to the early 1950s show an increase of 1.5°C in the mean annual temperature. Rainfall is down by 25 to 30 per cent over the same period. If you happen to be looking for a smoking gun, it's right here. Marion Island is being squeezed like a sponge by the fist of global warming.

South African scientists have been studying the island closely for the past half century; in fact, it has become something of a living laboratory. A small, weather-beaten research base on the eastern, leeward side of the island houses a team of meteorologists, biological scientists and support staff. Every year in March the *SA Agulhas* powers down into the treacherous Southern Ocean to collect the previous year's team, deliver a fresh group, and replenish the base's pantries. I was given a berth on the ship in 2003, following a team of researchers who were heading down expressly to look at how global warming is manifesting itself in that system. It is there that this journey into the realm of climate change began for me.

The Prince Edward Islands consist of two volcanic mounds, Prince Edward and Marion, that jut out of the southern Indian Ocean at 46° South, 37° East. Speaking of their isolation, author John Marsh said:

> Nature seems to have been at pains to maintain the exile of its orphan islands by creating between them and the inhabited lands of the Earth the barrier of the 'Roaring Forties'. In this zone that circles the Earth just to the north of the domain of the 'Westerlies', the wind is never still or constant in direction, and the sea invariably reflects its restlessness. Fog and ice add to the dangers that confront mariners venturing into such high latitudes. They keep well clear of them unless there is good reason to go that way.

The islands first made it onto the charts in 1663 during the early exploration forays into the Southern Ocean. Once the world's nautical frontiersmen began the race to find the last great continent, which geographers of the day said must cling to the underbelly of the planet – the 'anti' or counterbalance to the North Pole's Arctic, exploration of the region began in earnest. One captain, thinking he had finally discovered Antarctica in 1772, called the larger island *Ile de l'Esperance* (the Isle of Hope), and the smaller one nearby *Ile de la Caverne* (the Island of the Cave)[1]. Realising his mistake (and probably somewhat miffed that he may have missed his chance to discover the only remaining continent), the captain settled on *Ile des Friodes* (the Frigid Islands), and turned his offended back on them.

They later became the Prince Edward Islands, with the larger called Marion and the smaller Prince Edward, but much of the Southern Ocean's history was eclipsed by the excitement of Antarctic polar exploration. Nevertheless, there are a handful of tales about castaways, shipwrecks and occasional deaths as the world's demand for seal blubber, penguin eggs and guano kept people trekking down to this isolated, cold, wet little place.

It was only in 1948, after World War II, that South Africa laid claim to the islands. Weather forecasting in the Southern Hemisphere had become the hot topic as the scientific community had figured out that this was the world's weather factory. Post-war insecurity was another strong motivator. The development of long-range missile technology meant that possible enemies of the former Allied states could control substantial shipping routes around the southern tip of Africa, even attacking South Africa if they gained control of these islands and installed their missiles. In 1948, the Prince Edward Islands went from being a hellish little archipelago to really sought-after property. The South African navy's *SAS Transvaal* slipped out of Cape Town harbour late in December 1947 and headed south to annex the islands in *Operation Snoektown*.

Settlement at 46° 54'South and 37° 45'East was never going to be easy. The first visitors to Marion said:

> The island consisted of a lava rock base covered with a layer of decayed soggy vegetation many feet thick, with a carpet of luscious green Azorella and tussock grass as a top dressing. The ground was swampy and the going most arduous[2].

The mires are legendary. One of its first inhabitants described Marion as 'really a bog on top of a rock', while Marsh recalls someone sinking up to his neck in the rotting mud during the annexation. The first settlers were prohibited from walking beyond a four-mile radius of the new base because they believed it was simply too dangerous. A man – because only men went on the early ventures to the Prince Edward Islands – could easily become stuck in a mire and be unable to pull himself out if he became 'bogged in the muddy surface'.

Eventually, after several hurried trips by the South African navy to put buildings in place 'before the winter storms cut them off from contact, except by radio, for the best part of six months', a rudimentary base was open for business. It had meant struggling against 'gale upon gale, rain, sleet, snow and bitter cold', but the base had just enough room for twelve men and their meteorological and radio stations tucked away on the leeward side of the island.

Half a century later, the base is a rabbit warren of prefabricated buildings joined by raised catwalks. Since the first construction in 1948, it has been added to and replaced so many times that it is starting to look like an old shirt that has been mended too many times – eventually little of the original fabric remains and that which is there is faded and frayed at the edges. The buildings leak. The wind shakes them so ferociously that some people have complained of motion sickness during bad weather. Even the bunk beds sway when the westerly picks up. Some buildings have been blown away. For those who work here, far from the conveniences of the modern world, it is a small sacrifice to make to enable them to study the island's unique plants and all the strange animals that rest, nest and haul themselves out onto Marion's beaches. It's like living in a Petri dish.

If modern humans had been around a million years ago and had created the means to traverse the oceans, intrepid voyagers of the south could have sailed right through this point on the map without as much as a scrape to the keel of their boat. They would simply have found more of the Southern Ocean's unbroken and perpetually churning waters. But, at that time, our ancestors were just learning the nuances of stone tools and negotiating their way out of Africa and into Asia and parts of Europe. It took many generations before we populated most of the continents and figured out that the Earth was a globe around which vessels could be sailed.

Some time between 500 000 and one million years ago, the ocean floor below this point on the charts split open and bubbled with escaping gas and magma. Slowly, over the millennia, a submarine mountain range formed as layer upon layer of lava poured out and cooled itself over and over again. Eventually one of its peaks scaled the 4.8 $km^3$ above the sea floor, pushed through the ocean's surface and continued climbing. That, if you can imagine, is how Marion Island was born. A few hundred thousand years later, Prince Edward joined it.

Both islands started life above sea level as little more than a giant pumice stone – barren, cold and harassed by the elements. The first visitors were transient – seals and penguins arriving by sea, and albatrosses, gull-like skuas and petrels by air. Their stay on these inhospitable rocks would have been shortlived because, as yet, they offered little more than a place to catch their breath.

With time, and assisted by the passing visits of these travellers, the first immigrants made landfall. Plants arrived on the underbellies of birds or clinging to driftwood. Some, like pioneering *Azorella selago* and some mosses, colonised the nutrient-poor volcanic ash, scoria and rock despite their constant freeze-thaw cycles, ravaging winds and perennial rain. Other plants – mosses, lichens and grasses – would eventually arrive and also put down roots. Tiny creatures just visible to the naked eye – mites, springtails and weevils, and some not visible to the naked eye – like the moss-loving water bears (tardigrades), would have taken up residence in any habitable plant matter they could find. Over thousands of years, the natural growth and death cycles of these creatures, along with the constant raw edge of the elements' relentless pounding, enabled enough organic material to accumulate to plug up the porous rocks and allow fresh water to gather: the beginning of a sub-Antarctic garden.

Marion Island and its smaller sibling are geological newborns. Compared with ancient ecosystems on the mainland, these environments are gasping their first breaths and consist of little more than an apron of peaty mires around the coast, drier fern slopes, some grassy patches, and a polar desert further inland. To science, the beauty of Marion is beyond the obvious green, rolling hills of the lowlands and the wild, rocky interior. Its remote habitat and the simplicity of its ecosystem make it easier to understand than the much more complex workings of older ecosystems where many more tiers of life mingle in intricate and complicated ways.

Knowing all the plants and animals makes it easier to see where each prefers to live, as one climbs the gradients. Marion's lower slopes – from the coast up to about 500 metres above sea level – are covered with an undulating skirt of mossy mires, less moisture-dense fern slopes, and the occasional fell-field (wind desert), edged with a detail of frilly black lava rock and the vacillating spumes of green breakers. Lush, nitrogen-rich temperate grasses make comfortable homes for birds and mammals.

There are no roads or pathways on the island. Getting anywhere is by foot alone and usually dressed up in thermals, a waterproof outer layer, and industrial-strength gumboots. It's hard work stepping gingerly over mires, hoping not to punch a foot through the thin skin of vegetation into the belching mud below. It's like walking over a bowl of half-set jelly.

Dotted around like rusty pearls on fabric are iron-red volcanic cones. To summit these scoria mounds is like walking up a mountain of brick chips – two steps forward, one slipped back. Their upper slopes seem lifeless until you reach the craters, some of which are often full of water and fringed with green life. There are no trees or shrubs in this strange world.

Inland of the squelching coast is Marion's darker side. At about 500 metres above sea level, the island's skeleton protrudes – jagged and unforgiving – through strangely primitive plants such as mosses and liverworts. Young rock prickles from the ground in vicious, sharp barbs. Here and there the older weathered grey lava shows through younger deposits of rock that have tried to inundate everything over time. Above 800 metres, the only vegetation to survive is lichen and, at the summit (1 230 metres), there is virtually no plant life at all. Here, too, red cones bejewel the higher slopes. The desolation is breathtaking.

This place is home or occasional pit stop to the Southern Ocean's unique birdlife. The paddy (or lesser sheathbill) is the only land-based bird here and dashes about like a bantam hen, pecking at insects. A bird-spotter's trophy with its ice-white feathers and coal facemask, the paddy only occurs on a few sub-Antarctic islands. King penguins are the jesters of the islands, with their magnificent golden earmuffs and bibs, morning-jacket grey backs and ivory breasts. They waddle about, unperturbed by the intrusion of people on their beaches. They are very laid back, not having known any terrestrial predators. Even the sealers in the 1800s – who killed, skinned, boiled, ate and probably smoked anything that moved on the island – did not make enough of an impression on the animals to leave them skittish. They are also surprisingly curious and will tug and nibble at your clothing if you sit still long enough. Their cousins, the macaroni and rockhopper penguins, have comically plumed brow feathers which bob and dance like a tribal headdress. Rockies – not much bigger then rugby balls – find bouncing a far more efficient way of getting about than bipedal walking. With two sprung feet and little flippers, they navigate daringly up cliff faces with extraordinary deftness.

On Marion, the yarn goes, you get so close to nature that it includes you in its mating rituals. A radio operator once caused quite a stir when he visited a macaroni penguin colony. Resplendent in scuffed black, his steel-toed Wellingtons soon caught the attention of frustrated bachelors hanging round the periphery of the colony. Male macaroni penguins try to win a mate by dropping stones at her feet in a show of home-making prowess. Before long the radio operator had a cairn growing lovingly at his boots. Each bird twittered and twitched at his feet, plumed brow-feathers aquiver, appealing to the boots to declare their love.

Wandering albatrosses are born and bred for the Roaring Forties. Down here they harness the prevailing westerly winds with their impressive three metre wingspan and circumnavigate the globe in the high, southerly latitudes. These birds, like giant seagulls with strange dodo-esque heads, are believed to spend their formative years at sea. They don't appear to come in to land for as many as seven years or more in their youth and, when they do, they alight on the island of their birth. Here they will engage in an intricate courtship dance, pick a mate and remain faithful until death tears them apart. The life history of these extraordinary animals is only now being understood, even as their numbers dwindle on the sharp end of long-line fishing hooks.

Other albatrosses are found here too: the grey-headed, the Atlantic yellow-nose, the sooty and light-mantled albatrosses are all smaller but no less beautiful. Then there are the northern and southern giant petrels and their many cousins, as well as skuas, gulls, terns and the skittish gentoo penguin.

There's another bird that should, if it cared about such things, bemoan the injustices of social hierarchy. The Crozet shag used to be called the imperial cormorant before it was renamed.

'That's like being demoted from royalty to the village tart,' one scientist observed to the backdrop of squawking shags.

The largest mammal found here is the southern elephant seal, an animal of enormous bulk that can swim to extraordinary depths. Scuba divers stay above thirty metre depths unless they have back-up. South African extreme diver Nuno Gomes achieved a new world record by diving to 318.25 metres in the Red Sea in June 2005 – it took him 20 minutes to reach that depth and 12 hours of laborious decompression to reach the surface again. The deepest free dive has reached 171 metres. Elephant seals go down more than 1 500 metres. Between dives, the animals will ventilate their lungs for about three minutes, breath *out* and then head down into the deep where they can stay for as much as two hours on a single exhaled breath. They have more haemoglobin in their blood and myoglobin in their muscles compared with humans and a relatively high volume of blood. With more places to store it, the animals can bank plenty of oxygen and survive between breaths for these improbable lengths.

Pressure increases by one atmosphere every ten metres under water. At 1.5 km down, therefore, the pressure is 150 times that of the surface[4]. To survive this crushing pressure, elephant seals have evolved without any cranial sinuses; their lungs and ribcages are collapsible, while their incompressible muscles and bones consist of 85 per cent water, which counterbalances the immense pressure. With no oxygen in their lungs and a system that does not allow gas exchange, the seals do not experience the problem of nitrogen gas bubbling in the blood stream when they resurface. This means they are spared the pain, convulsions and loss of consciousness associated with the 'bends' that human divers suffer if they surface too rapidly. Elephant seals can also tolerate higher levels of nitrogen, saving them from nitrogen narcosis[5].

Antarctic and sub-Antarctic fur seals have deceptively cute puppy-dog eyes. But they easily charge in a volley of emphysemic, rattling coughs. Their bite can slice effortlessly through gumboots and, if they hit flesh, can bring on sepsis in hours. Further down this food chain are all the smaller animals – weevils, the anachronistic flightless fly (which, if it could fly, would be blown out to sea), beetles, moths, a single wasp and gnat, springtails, tardigrades, a midge species and a handful of other tiny beasts. All these creatures – from the 3.5 ton male elephant seal down to the microscopic water bear – have inextricably woven themselves over thousands of years into a complicated food web where, ultimately, every species influences another in some way.

The name 'sub-Antarctic' conjures images of frozen wastes, psychedelic night-lights, twenty-four hour summer days and icebergs. On the contrary, this part of the world is comparatively mild.

What makes it unique is that it is relatively constant. Typically the island is pummelled by strong westerly winds all year round. Conditions are usually overcast (bright sunshine is rare) with torrents of year-round rainfall. Occasionally a deposit of snow and sleet will occur.

There is little variation in temperature. While summer may bring a 22°C high (that's about the record high) or minus 6°C in winter, the mean annual temperature is about 5°C. Summer's average is 7.3°C and winter's average is 3.2°C.

Scientists get excited about a system like this. They have a full inventory of plants and animals (including non-indigenous species that have been introduced over the centuries) and comprehensive weather records dating back over the past fifty years. It's as close to a controlled experiment as nature is going to give us. All this makes Marion Island a living laboratory for the study of climate change in the Southern Hemisphere. The ice plateau was never formally mapped, but an estimated twenty to thirty hectares of ice covered the interior. A geomorphologist who visited the island speculated that, if the plateau were at an equilibrium thirty years ago, it is possible for that amount of ice to have melted below the surface with just a 1.5°C mean warming.

Marsh called the island a jade jewel fifty years ago. Now it is a jaded jewel. Over the centuries that humans have been visiting these shores, stowaways have come with them, hidden in cargo, clothing, food and the bowels of ships. Plants, insects and mice – alien to this ecosystem – were inadvertently introduced. For decades the cold, wet, windy climate kept their numbers in check. Growth of these foreign species did not become too invasive, but this is changing with the steady warming and drying over the past fifty years.

Dr Niek Gremmen, a specialist in sub-Antarctic botany, has been studying the island's mosses and other plants since the 1960s and has kept a careful eye on the movements of the exotic plants introduced by humans. *Sagina procumbens,* a moss-like plant, was first identified in the vicinity of the research base in 1965. Back then it didn't worry researchers. Today Gremmen paints a very different picture. Between 1965 and 1975 invasive species spread by 30 to 100 metres per year on Marion Island. In the past decade this has increased to 200 metres per year. Now *Sagina* is spreading at a rate of 320 to 370 metres per year in certain places on the island. The plant was first noticed in small pockets on nearby Prince Edward Island in 1997, probably taken there by birds flying the 20 km distance between the two islands. Gremmen, who spent six days on Prince Edward in 2003, says *Sagina* is spreading at a rate of 600 metres per year in places.

The botanist has small research plots on Marion that he has been monitoring for years. Comparing plots of natural vegetation with those containing invasive species, Gremmen has observed that there are 50 per cent fewer indigenous species in the invaded plots. As the climate gets warmer and drier, the conditions become more conducive to the survival of the alien plants, which hail from warmer and drier climes. Increasingly the aliens are inundating their more timid counterparts.

A fierce but small-scale battle rages among the animals here, too. In its natural state, Marion's residents gravitate to their own exclusive suburbs. Seals and penguins tend to stay close to the shoreline. Comfortable beaches are at a premium and that's where you will find them. Occasional seal wallows, fetid with stale water and excrement, are testimony to the nutrients they unload every time they haul out for a break from fishing. Penguin guano comes with its own nasal insult. There is plenty of fertilizer here and the plants thrive in it. When the rains come – which they frequently do – much of this is washed back into the ocean where the algae bloom spectacularly. The zooplankton then gorge on this marine 'pasture'; fish fatten themselves on the zooplankton; and the near-shore feeding penguins feast until their numbers are bursting at the seams. It's a reliable, self-perpetuating engine.

Drier inland grass areas are difficult to reach for the swimming types. Vegetation here depends on ocean-faring birds to fly in any nutrients when they come to nest. The mires are more exclusive – what science calls a 'closed system'. They resemble sludgy puddles and, even with a fragile mossy cap over the mud, they fail to invite nesting birds, so little nutrient supply is imported to the mires. The mires must therefore feed on themselves: the plants grow by absorbing carbon dioxide from the air, some nutrients from the slushy ground beneath, and taking in a bit of sunlight. From this they construct their stems and leaves. If the plants were to die without passing through the gut of a herbivore – which would release the nutrients back into the soil for the plants to forage on – the dead plant material would simply accumulate in a layer of 'litter', sink down into the marsh and, over time, compact into a layer of peat. Add a few hundred million years and coal would form.

Without a constant fresh delivery of nutrients, the soil would eventually run out of the good stuff such as nitrogen and phosphorus and exhaust itself, in much the same way as a pot plant would slowly expire if its soil were not replaced or topped up with fertilizer. Slowly the mires would become worn out, sterile puddles of sludge. But they are not barren, which clearly suggests that something else is at work here. Driving this engine onwards are the hungry mouths of a host of diminutive creatures: caterpillars and earthworms graze their way through the detritus of these mires, feeding on the litter of dead leaves, recycling the nutrients back into the soil and keeping the live plants healthy, green and wobbling over their muddy substructure.

This system has taken several hundred thousand years to reach a self-perpetuating equilibrium. In just fifty years it is being thrown out of kilter. The first import to noticeably disrupt the system was the house mouse, which arrived on the island with sealers in the 1800s.

Mice feed on grass seeds and invertebrates – insects, arthropods and spiders – living in the vegetation. Even outside the mires, the growth, death and foraging cycle of these invertebrates make an important contribution to the nitrogen requirements of some of Marion's plant communities. They also sustain the lesser sheathbill. Stellenbosch University botanist Professor Valdon Smith has been studying Marion for three decades. He says a fifty per cent reduction in the sheathbill population has been recorded since 1976. A lack of food – invertebrates – owing to competition from mice is the most likely culprit. This compares starkly with Prince Edward Island, which has a stable sheathbill population and no mice.

Mice have made themselves at home across the island – mainly in the tussock grass – from the coast up to as high as the 750 metre contour. They thrive in summer and then die back in winter as the chill culls their numbers. Without natural predators, it falls to the sub-Antarctic climate to keep the mouse population in check. Yet Smith says an increase in their numbers has been observed since the sixties. As conditions become warmer and more favourable to their survival, their population is expected to boom at the expense of the locals.

After annexation, mice became such a pest that some of the first settlers shaved their heads – in spite of the cold – because the mice kept nibbling at their hair at night. In a flush of good intention, five cats – a single female, a neutered male and later three kittens – were brought to Marion in 1949 to tackle the mouse blight.

Within two years, the cats' offspring had gone feral. By 1975, 2 000 cats roamed the island, by which time they had discovered that birds were far easier prey than little rodents hidden in the undergrowth. The island's burrowing bird population was hardest hit. In that year alone, the cats managed to eat 'just under half a million birds, resulting in the extermination of the common diving petrel and the near extinction of three other species (of petrel)'[6].

By the 1980s it was time to take on the cat population. It took five years, innumerable man-hours, traps and nocturnal hunts but, by 1991, the last cat had fallen. Some twelve years later the burrowing petrels are returning, bringing with them the nitrogen from their guano that supports so much plant life. In April 2003 Gremmen and Smith noted an increase in the amount of tussock grass found near the coast. Good news, they confirm, because it indicates a recovering ecosystem. Even so, the bird numbers pale compared with Prince Edward Island, which never lost birds to the unwelcome felines. There the burrowing petrel communities are so strong that Gremmen and some colleagues who enjoyed a rare visit to the island in 2003 had trouble sleeping at night because the vocal night birds were so numerous.

Many of the plants and animals on Marion occur nowhere else in the world. As the climate changes and alien species flourish, indigenous plants will be out-competed and extinctions will occur. Like extinctions, states Gremmen, the introductions of alien species are forever. You can disrupt an area and nature will recover, even if it takes fifty years or five hundred. But if you introduce an alien species there is generally no going back.

Now that the cats have been removed, and the impact of the mice has been understood, and as fears of global warming grow, South African researchers have realised the value of Marion Island for mapping the impact of climate change on plant and animal communities.

Local species have evolved in extremely chilly conditions, so everything about them seems to be on a go-slow. While a house fly in South Africa will hurry through a full life cycle – emerging from an egg to becoming an egg-laying adult – in just four days, Marion's kelp fly will take six months to a year to get through the same stages of life. A weevil takes a year, which is much slower than a warmer climate weevil. Marion's flightless moth, *Pringleophaga marioni*, also trundles through its lifetime. A female *Pringleophaga* moth will lay forty to fifty eggs that will hatch after about a month, which is a relatively slow incubation period for moths. When the caterpillars emerge, they will set upon local vegetation with relish – and continue to do so for the next four to five years. Only then will they pupate and set about the business of becoming adults.

Invasive species are high-octane breeders. Having evolved in warmer climes, yet still able to withstand the southern chill of Marion, they rush through their life cycles. The exotic springtails, cabbage moths and blow flies cover three to four times the number of generations per year than their indigenous counterparts. More significant, however, than the relative speed – or lack of it – of the exotics versus the locals, is their response to increasing temperatures. Exotic species, already on the gallop, respond quickly to an increase in temperature and so get through even more generations than before. Indigenous species respond slowly to temperature increase.

Consider exponential increases and you will get the picture. Before long, Marion will literally be crawling with foreigners. In the long term it means there may well be more litter recycling and nutrient production – signifying more food for the plants. But huge increases in the number of alien species will reduce the amount of peat formed in the mires, challenge the indigenous species at their own game, and entirely change the functioning of the system.

Meanwhile the penguins face their own uncertain future. The ocean currents around Marion ebb and flow to their own unfathomable rhythm. South of this area, around the Antarctic continent, is the Antarctic Polar Frontal Zone. North is the Sub-Tropical Convergence. Both are massively powerful oceanographic fronts. Running between them, and directly in the path of Marion, is the weaker Sub-Antarctic Front. It seems that the ocean immediately around the Prince Edward Islands swings between a two-phase system. In some years strong currents push between the two islands, bringing with them organisms from the deep ocean. This makes excellent feeding for the flying birds such as the petrels and albatrosses, which can reach the offshore food. When the system switches to the second phase, the currents do not push between the islands. All the nutrients that are washed out from the penguin and seal colonies accumulate close to the shore instead of being taken out to sea, bringing about the massive algal blooms mentioned earlier.

Near-shore feeders – the penguins that swim for their food – relish in an orgy of feasting. But the algal blooms along the coastline of Marion are on the decline. The Sub-Antarctic Front has shifted south relative to the islands by as much as 1.5 degrees since 1974, making way for warmer waters from the north. The temperature of the ocean in the vicinity of the island has crept up steadily by 1.4°C over the last fifty years of records, in parallel with the 1.5°C increase of Marion's mean temperature. The fate of the rockhopper in these parts depends on where this trend is headed. A massive 94 per cent decline in rockhoppers has been recorded at Campbell Islands near New Zealand since the 1940s. While the study that reported this decline omitted over-fishing as a possible factor, it did state the decline was associated with rises in sea surface temperatures and reductions in the penguin's fish source[7].

Conservation ecologist at Stellenbosch University, Professor Steven Chown, has researched the islands for years and says the question remains whether this is an ongoing trend or a temporary anomaly. 'Will the Sub-Antarctic Front end up on Antarctica's shores? That seems a bit unlikely because the Antarctic Polar Frontal Zone has been stable there for the past thirty million years,' comments Chown. 'But the other question is whether there's been a regime shift. So instead of this linear trend of continuous warming, who knows where it's going to end? We may have seen a shift from one system to another, which may be about to stabilise. We don't have enough data to know for sure. We haven't recorded for long enough to know whether that's happened.'

*Azorella selago*, a curious cushion-like plant, has already assisted scientists in understanding how climate change will shape the plant life of the island. It is known that *Azorella* helped to pioneer life on Marion Island – it is a tough, slow-growing plant that forms in a tight configuration with hardy stems and leaves. This makes it able to withstand the extremes of polar deserts and the island's near-barren lava slopes. But it doesn't like to compete for real estate so, on the lower, lush contours of the island, it is easily overrun by grasses, ferns and moss.

*Azorella* moves in where few others can go, establishes itself, and sets up comfortable little habitats where communities of mites and other insect-like creatures can live. These creatures are nineteen times more prolific in the cushions that on the fell-fields surrounding them. *Azorella* is a pleasure dome to life forms that simply could not hack it on Marion otherwise. Many *Azorella* cushions have grasses waving out of them like ostrich feathers from a hat – the plant allows the less robust grasses to take root in areas that would otherwise be too hostile for their survival. Where *Azorella* leads, other life forms follow.

A series of experiments have been conducted on some *Azorella* plants near the research base. A community of plants on a fell-field to the north-east of the island was exposed to a simulation of conditions that might be expected if climate change predictions come to pass. Some cushions were sheltered from rain and allowed to warm. Others were shaded. After a year, their condition was compared with non-manipulated cushions. Associate professor in Stellenbosch University's Department of Conservation Ecology, Melodie McGeoch, explains why *Azorella* makes a good case study. 'Azorella is unique because it grows on fell-fields in a type of monoculture. It forms islands, each with its own discrete boundary, which makes it so much easier to study. These islands mean there are few influencing factors so it's easier for biologists to understand than a complex, open-ended system.'

It's like a metaphor for Marion's greater ecosystem. 'If you get to the bottom of how this system functions, then potentially you can extrapolate to other systems.' Most of the island's indigenous and alien plants occur in its more hospitable lower altitudes where *Azorella* struggles to survive the throng. If the climate continues to warm and dry, these plants will move higher up the island as conditions become more favourable and directly into previously inhospitable areas where *Azorella* enjoys a roomy, if lonely existence. These species may overwhelm the cushions.

*Azorella* will also move higher and further into the hinterland, trailblazing as it goes.

The experiment findings show that shading – a simulation of the secondary effects of climate change as grasses inundate the cushions – causes etiolation, where the stems grow further apart and the leaves bigger, making them less resistant to ice and wind. Reduced rainfall produced stem damage in the experimental plants, with earlier autumnal aging and dying back.

McGeoch concluded that climate change – in terms of decreased rainfall and the indirect effect of shading by other successful plants – is likely to take its toll on *Azorella*. Since it is a nutrient-rich life support to others, the impact will ripple through this fell-field ecosystem.

Studies of the responses of the cushions' biological inhabitants – about thirty species of mostly springtails and mites common to the Antarctic – supported the experiment team's conclusions. The number of tiny animals present in cushions dropped radically as their home dried out. But each species responded uniquely to its changing climatic conditions. Some thrived while others disappeared. It is impossible to say whether species packed up and left the cushions when things became too dry for them, or whether they died, or were simply less able to reproduce successfully. Whatever the case, it's safe to say that even in as 'simple' an ecosystem as an *Azorella* cushion, there is no quick answer to how an entire suite of species will respond to the warming, drying and other changes expected to impinge on them as global climate change gears up.

In the meantime, strategies are in place to save the ecosystems of the Prince Edward Islands from further ruination. A strict environmental management plan controls who visits the islands and how often. The plan also does whatever is possible to prevent foreign plant or animal species from sneaking in with every visit. Prior to the departure of any ship bound for Marion Island, conservation officers inspect the ship and all equipment and food containers for insects, seeds, mice or rats. Before disembarking from the ship, all people visiting the island have their boots scrubbed with bleach in a ritual boot-washing ceremony. Backpacks and any other gear going out into the field are inspected. Fresh fruit and vegetables, seeds or pips of any kind are not allowed anywhere near the island. Even poultry products are irradiated to prevent the introduction of avian diseases to the island's birds. Poultry waste is frozen and returned to the mainland for disposal.

The changes that have already taken place on Marion are symptomatic of a broader problem – of a planet that is becoming 'a big corporate world that is homogenous from north to south, from east to west, dominated by a few weedy feral species,' comments McGeoch.

'We are such a widespread species, with a highly effective transport and communication system, that we are introducing the same species all over the world. Eventually we are going to be living in a homogenous environment.'

Do we want this, McGeoch asks, or do we want to 'live on a planet that is diverse, where every part looks unique and operates differently, where a variety of species are interesting and look different'?

Chown explains it like this: you wouldn't want a single, homogenous system for the same reason that wise investors would never plough all their capital into one policy because, if it folds, they are history. For the planet to absorb changes, we are going to need a bit of variety. Biodiversity is the blueprint of the future, whatever that turns out to be. Allowing species to die is like plundering your savings.

In medical terms alone, nature provided the world with nearly $200 billion worth of pharmaceuticals in 2002 – half the global expenditure for health cures in that year. Biological diversity is described as the sum total of 1.75 million species of plants, animals and micro-organisms (and the many more not yet discovered), as well as the systems in which they struggle with one another in search of their own individual niche in the greater scheme of things. This vast wealth, which has been banked away during billions of years worth of evolutionary process, allows the human species to continue sifting through nature's secret and prolific stash to find cures for their prevailing physical and psychological ills. And this is just a fraction of the greater number of goods and services that nature provides us with daily.

'That's the economic reason,' Chown states. 'The philosophical reason is straightforward. Are you supposed to look after things or just kill everything?'

It's not a question you can resolve easily, he comments. Every person may have a different opinion on this but there is no doubt that people are really short-sighted if they think they can live without the biosphere around them. Where would they find their food? Destroying the biosphere, Chown says, is the 'equivalent of going out and just shooting every species you see – simple as that'.

Affording the islands this degree of protection means a policy of not allowing tourists to visit the reserve. Not everyone likes it, but it's probably the only way to prevent further damage.

So, when a pilot crash-landed his plane on Marion Island in November 2002, many of the island's scientists were furious that one person's foolish attempt at personal conquest put the island ecosystem at risk.

The French pilot, Henri Chorosz, was attempting to circumnavigate the globe via the poles in a home-assembled kit aircraft not much bigger than a shoebox. As he approached the higher southern latitudes, ice began to form on the aircraft's wings, making the machine guzzle its fuel load. Soon Chorosz realised that he might have to ditch in the Southern Ocean where he would surely die. Instead he found this tiny dot of land conveniently nestled in the Southern Ocean. At 5 am on a sunny, austral spring day, residents of the island saw the light aircraft appear over the horizon. Turning into the wind, the pilot managed to make an emergency landing as chunks of ice fell from the aircraft's wings. The wheels touched down in a near perfect landing – until they dug into the deep mire. His plane, the *Little Transpacific*, performed a perfect head-over-heels and came to rest upside down. Once the medic had finished checking him over, Chorosz was set upon by a conservation officer. His shoes were doused with insecticide and herbicide and the aircraft's undercarriage was treated to a similar welcome ceremony. It's not that they wanted to be inhumane, explained the team leader later, but they had Marion to look after too.

# Chapter 13

The 'Green Revolution' has not meant equal food surpluses for all
and now climate change will hit Africa hardest.
Photograph © Guy Stubbs/Independent Contributors/africanpictures.net

# Food for the human animal

*The satiated man and the hungry one do not see the same thing when they look upon a loaf of bread.*

Rumi, poet and mystic, 1207–1273

How many of us will ever know real hunger? Hunger of the kind that swallows your belly and claws at the spine, or the lethargy and depression of prolonged malnutrition. Somewhere beyond the comfort of so many of our modern lives, 1.1 billion people live with hunger as a perpetual houseguest as they scrabble for the basics of survival on less than $1 a day.

The closest that I have come to making hunger's acquaintance is a trivial and frivolous anecdote from a fruit-picking season in the United Kingdom. I was between jobs on one of those indulgent journeys of self-discovery. You know, the kind where middle-class kids avail themselves to the rank and file of manual labour to fund a bit of Euro-travel. The kind involving hours of sitting on a backpack around stations, smoking roll-ups and contemplating the meaning of life. What a thoroughly bourgeois thing to do.

I was about to learn what happens when you run out of cash far from home. A train spat me out into a downpour on a Kent station, far from the hops farm where

I needed to be. Slumping into the rain, I hitched my backpack higher on my shoulder and clutched a bag of sodden groceries. My pockets hung limp with a single pouch of half-empty Golden Virginia, disintegrating Rizlas and about £2-something in loose change. There was not enough to get through ten hours of manual labour a day until pay day at the end of the week and it stirred a most primal insecurity.

Starvation has prowled just beyond the light of the campfire and stalked through our settlements as long as humanity's ancestors have been walking upright. The Vikings, for one, had an unfortunate encounter with it. Long before Christopher Columbus stole the limelight, they had already attempted to open up North America to trans-Atlantic trade[1]. These intrepid Nordic seafarers were aggressive colonisers during the peak of their culture – from about 800 to 1100 CE (Common Era). They set sail from their homeland of present-day Norway and Sweden and settled in Iceland, Greenland, parts of Scotland and Ireland, and even travelled as far as Newfoundland.

Their success was aided by a particularly balmy spell of post-glacial weather. Since the retreat of the continental ice sheets about 12 000 years ago, the planet has enjoyed a few comfortable warm spells. The Medieval Warm Period between 900 to 1200 CE[2] was one such time when temperatures were, on average, 1°C warmer than they are today. It was a pleasant time – pack ice retreated northwards, opening up safe travel for Viking ships, and longer growing seasons meant productive agriculture. Viking farmers grew cereal crops in Iceland and Greenland and had enough surplus grain to sustain their livestock through the frigid winter months. And they traded with other parts of Europe for the items that they were unable to farm, their ships criss-crossing the Labrador and Norwegian Seas.

Then the Little Ice Age happened. Most argue that a decrease in sun spot activity cast the Northern Hemisphere into a chill. The impact on Europe was dramatic. A 1°C decrease in average temperatures cut the wheat-growing season by a third in Iceland[3], reducing cereal surpluses and making it difficult to keep cattle alive through winter. Once pack ice began drifting south again, Greenland was cut off from its European trade partners.

The strapping Nordic Vikings, who at the height of their achievement averaged 1.7 metres tall[4] (which was pretty tall for the time), shrunk to under 1.52 metres tall[5]. They were forced to abandon cereals, livestock died and they were left to rely solely on the sea for nutrition. Their population decline was dramatic, too. The last Vikings in Greenland were 'severely crippled, dwarf-like, twisted and diseased'[6].

Let's go back a little further, to about 10 000 BCE (Before Common Era), when humanity was embarking on its first tentative steps into property investment and the security of white picket fences. By the time Northern Hemisphere ice began to retreat at the end of the most recent glacial period, we had long mastered our opposable thumbs. Our handiwork with rudimentary tools set us apart from other mammals and our primate cousins. We had mastered the art of walking upright. Our voice box had dropped lower in the throat, positioning the larynx in a place that allowed for a new kind of communication, which served well to manage the hunt and to co-ordinate groups. In three million years our brain had grown to three times the size of our earlier australopithecine ancestors[7].

As hunter-gatherers, we were the wandering type, living comfortably off the land. Recent studies of present-day hunter-gathers suggest that we probably had ample leisure time in our pre-farming days – adult !Kung-San Bushmen in the Kalahari only need a two-day work week to keep their families in good health[8]. We had found our way onto every continental landmass and were happy drifters, adapting to the environment with our smart, well-organised brains. Except for the danger of the occasional hunt, food was easy to come by.

So why on earth did we abandon these bohemian ways to settle down into domesticity? Pursuing an agricultural life, in truth, is more physically demanding. It also opened us up to the vagaries of nature where pests, disease and capricious weather could be dangerous to groups who had lost the ability to pack up and leave when things got difficult. Several explanations have been offered for why we chose to settle. One theory suggests that, at a global population level of four million, competition for resources was increasing. We were running out of new places to exploit and had to find more intensive ways to obtain our required daily calory intake. Some theorise that the advent of pottery made it possible to store food better, which made it worthwhile to adopt the kind of lifestyle that would generate a food surplus[9]. Waste not, want not!

A more recent idea hypothesises that agriculture was a natural evolution of our relationship with plants and animals. Slowly we modified our environment and encouraged the plants we preferred to eat, giving a little water here or burning some areas there, and frequenting the places that we knew had the food we wanted. We continued to hunt and gather but slowly we began taming the land and domesticating animals which, typically, were 'placid, slow-moving . . . ate a wide variety of food and existed in highly social groups with submissive herd structures'[10]. When we set up camp we probably began to linger in places. Once the transition was made from drifter to homemaker, there was no going back. It's called the ratchet effect: the ability to produce more food made the population thrive and locked us into intensive farming. It was impossible to return to 'less intensive ways of gaining food'[11].

Ingenuity in the field – irrigation, the plough, animal power – produced food surpluses that fed a growing population and allowed some people with a different bent to pursue activities other than running the family farm. These people became wheelwrights, carpenters and shoemakers. Soon villages sprung up with shopkeepers, politicians, middle managers and war-mongers – welcome to the earliest onset of civilisation. Whatever the reason for choosing domesticity, it tied us into a fragile alliance with the land.

Since then our survival has leaned heavily on that thin margin of food surplus, mostly of cereal crops. We need them for our morning bowl of porridge as well as to feed our livestock, whose roaming days have been tethered by our own sedentary ways. If crops fail for more than one successive season, the results could be disastrous.

• • • • • • •

Apparently the human body is quite resilient. Cutting our daily calorie intake by half will only decrease the body weight by one quarter. But reducing it beyond that threshold is dangerous – first the body burns up its fat stores before it dips into the protein pantry in our muscles and organs. Then the immune system weakens and we are highly vulnerable to disease. Children under five are least likely to survive.

Roughly four hundred major famines have been recorded through human history. I stumbled upon a few spectacularly grisly ones. In parts of Europe, unusual amounts of rain fell between 1315 and 1317, submerging many crops. Desperately needing to fill their aching bellies, farmers slowly began eating into the seed for next season's plantings. The food supply dropped and prices increased. Society was in crisis:

> (W)heat prices tripled, shortage rose eight-fold. The poor could not buy food and in some places there was simply no food available. The poor were dying in huge numbers . . . The food that was available was of very low quality – bread contained pigeon and pig droppings and animals that had died of disease were eaten, causing outbreaks of disease among humans. There are widespread reports of cannibalism from Britain to the Baltic; in Ireland in 1318 bodies were dug up from graves to provide food and in Silesia the bodies of executed criminals were eaten. Lack of fodder and numerous diseases killed over two-thirds of the sheep in some areas and in the four years after 1319 about two-thirds of Europe's oxen died.[12]

A hundred years earlier, at about the time when the Vikings were trying to negotiate their way through their own crisis in the shape of a mini-ice age, an El Niño event swept crippling drought through the Middle East. A medieval Iraqi physician and scholar, Abd al-Latif alBaghdadhi, gives an account of the famine that hit Egypt in 1200 and 1201 CE[13].

Travelling through the region, on one occasion he saw 'a mountain of over 20 000 corpses' and on another saw a road which was like 'a vast field sown with human corpses'[14]. Over one hundred thousand people were buried in Cairo, but many more never found a grave.

People went to horrific lengths to survive:

> Cannibalism became so rampant that those walking the streets were caught

by hooks let down from windows above, tradesmen were lured to fictitious jobs only to be slaughtered while they worked, physicians consumed their patients, and hosts devoured their own friends after inviting them to dine.

Abd al-Latif . . . [saw] 'a small roasted child in a basket. They carried it to the Emir and led in at the same time the mother and father of the child. The Emir sentenced both of them to be burnt alive.' And even the punishment of death by fire did not deter the masses. The body of one cannibal, put on public display after being burnt for his crime, was found, 'Devoured the next day: the people ate it very willingly, for the flesh having been well roasted did not need further cooking[15].'

· · · · · · ·

The Irish learned the hard way that monocropping is one of farming's more risky practices. Struggling under a punitive rent and taxation system imposed by their British colonisers, subsistence farmers turned to potatoes, which produced relatively large yields from small and soil-deficient plots. In 1845 the potato blight found its way from America into the Irish potato crop. When the crop failed across the island, a million lives were lost – probably one eighth of the population. Many more emigrated to Britain, the United States and Canada.

In India, failure of the monsoon claimed 5.5 million lives (nearly double the population of present-day Cape Town)[16] from 1877 to 1879[17]. At the same time, drought across northern and central China starved nine to thirteen million people to death – the worst natural disaster in history[18].

The tragedy about famines and their gargantuan mortality rates is that they are not always caused by a shortage of food but by 'lack of available land, poverty, rising food prices . . . leaving people either unable to grow their own food to survive or earn enough money to buy it'[19]. Lack of political will has claimed its share of decimated corpses. During the Ethiopian famine of 1984, when nearly one million people starved to death, the Western girth was straining at its waistband. Europe had a bumper harvest that year and ample food made its way to the struggling African Sahel region. Ethiopia's government chose to reroute that food from its starving poor to the troops fighting a border war with Eritrea[20].

Similar political stupidity, evident in the collectivisation of agriculture by the Soviet Union's Bolshevik government, resulted in a food shortage that claimed the lives of seven million people in 1931 and 1932[21]. The same collectivisation policies implemented by Mao Tse-Tung's regime in communist China brought about the worst famine in human history. Joseph Becker's text *Hungry Ghosts, Mao's Silent Famine*, states that the rest of the world struggled to believe the figures which finally emerged – 30 to 40 million people are believed to have died from 1958 to 1961. That's nearly the population of South Africa, wiped out in a matter of three years. Many of those people are thought to have been young girls, allowed to die because of their low status in Chinese culture at the time[22].

I will probably be accused of cheap scaremongering. But the urbanised First World is so far removed from that tenuous link with the land that perhaps a bit of a shake-up once in a while would not go amiss. We stroll through supermarkets and forget that the neatly packaged lamb chops we pop into our baskets were part of an animal that had to eat grain and be bloodily slaughtered before it could be presented, neat and clean, on our carving boards. We stock up our pantries, oblivious to how far food has travelled or the soil from which it comes. We don't even have to wash our potatoes or chop our garlic these days. We stock our trolleys with Cabernet-blends from Napa, cantaloupes from Israel, beef from Argentina, coffee from Brazil, lamb from the Karoo, salmon from Scotland . . . and cannot fathom a world where none of that abundance is laid out for us each day down at the local grocery store.

Ancient famines were localised and took place at a time before air travel and modern communication made it possible to ship food from lands of plenty to places where babies no longer had the energy to suck from their mothers' breasts. Surely our great advances in modern transportation and global communication must come to the rescue of every skeletal child at the short end of decades of food shortage and war? Surely the Green Revolution will save the world's hungry? Thanks to advances since the 1960s in high-yield crops, irrigation, fertilizers, pesticides as well as other improvements in farm management, food production is keeping pace with population growth[23]. Now that the global population has broken through the six billion mark, we've got nearly half of the land's surface under agriculture[24] and there is more than enough food to go around.

It is true. There's more food to go around these days and even poorer countries are eating better. The United Nations Food and Agriculture Organization (FAO) reports that the percentage of undernourished people has been dropping since the late 1960s and the average daily calorie intake of developing countries is expected to 'exceed 3 000 kilocalories (kcal) by 2030'. Diets will become more varied for more people and nutrition will improve.

Globally speaking, that will happen. But the undernourished persist. Worldwide, the FAO estimates that 842 million people were undernourished in 1999 to 2001 – 10 million of whom were in industrialised countries, 34 million in countries in transition, and 798 million in developing countries. 'The numbers of undernourished continue to rise in sub-Saharan Africa and in the Near East and North Africa.'[25]

The United Nations Millennium Ecosystem Assessment published early in 2005 claims that more than two billion people still live in the dry regions of the world. They:

> Suffer more than any other part of the population from problems such as malnutrition, infant mortality, and diseases related to contaminated or insufficient water. Areas such as sub-Saharan Africa are among those where natural services are most threatened by human impacts. Bucking the trend of the rest of the world, the amount of food produced for each person living in this region has actually been going down.[26]

By 2015, six per cent of the world population (412 million people) will eat less than 2 200 kcal per day while sub-Saharan Africa will be the most undernourished.

No matter how great human advances, they have not spread current food supplies evenly across the world like a neatly buttered slice of toast. The second half of the twentieth century has produced enough food to fill every stomach on the planet. But, towards the close of the century, most of it was being eaten by western Europe, Japan and North America where 'half the world's food was eaten by only a quarter of the world's population'[27]. It didn't help that the hungrier nations were producing crops for export to ensure the industrial countries had a bigger and more varied diet. During this time, the 'domestic cat in the United States ate more meat than most people living in Africa and Latin America', writes historian Clive Ponting.

So while many in the First World literally eat themselves into an early grave with adult-onset diabetes, heart disease and other obesity-related illnesses, elsewhere in the world a child dies from hunger-related causes every eight seconds[28]. Sub-Saharan Africa, after decades of post-colonial strife, war and governmental mismanagement, remains one of the most nutritionally-deficient regions of the world[29]. Mostly, states the United Nations, because of 'its lack of political will to tackle poverty'[30].

In March 2003:

> Well over 15 million people in Zimbabwe, Zambia, Malawi, Lesotho, Swaziland and Mozambique faced immediate shortages [due to a] combination of unfavourable weather conditions, specifically droughts and flooding; poor macroeconomic management; and political turmoil, result[ing] in the worst regional food security crisis in the past decade[31].

Drought has been the mainstay behind food insecurity and famine in Africa during the past three decades[32], which does not bode well for the region in a changing climate. While the global trend for per capita food production goes up elsewhere, across Africa it has been declining for the past two decades. This, combined with the fact that more than half of the continent's population subsists off the land, has resulted in 'widespread malnutrition, a recurrent need for emergency food aid, and increasing dependence on food grown outside the region'. The World Resources Institute reported in 1998 that food consumption across the region exceeded production by 50 per cent in the drought-prone mid-1980s and by 30 per cent in the mid-1990s.

Now add climate change to the equation. No matter how good all the technical advances brought by the Green Revolution, food production is still dependent on water, soil moisture, nutrients and temperature – at least three of which factors will change in the near future. A change of up to 1°C could result in greater productivity at the higher latitudes globally, depending on the 'crop type, growing season, changes in temperature regimes and seasonality of precipitation'. But crops grown in the tropics and subtropics are already close to their maximum temperature tolerances – a temperature increase of just 1°C will compromise their annual yield. An increase of more than 2.5°C would increase world food prices. This amount of warming would also diminish any improved productivity experienced in the middle to high latitudes while the tropics and subtropics would see a more severe decrease in productivity[33].

> Undernutrition is a fundamental cause of stunted physical and intellectual development in children, low productivity in adults, and susceptibility to infectious disease in everyone. Decreases in food production and increases in food prices associated with climate change would increase the number of undernourished people.[34]

Most of this creeping hunger will be in the developing world. South Africa is precariously arid, but shifts unexpectedly between extremes of drought or flood, with rain falling erratically in different regions and between years. Even without climate change, the country is expected to use up its available surface water by 2030[35]. With climate change, rainfall will become even more unpredictable, possibly changing in its intensity as well as the seasons in which it falls[36]. This will, of course, reflect in the amount of water running off into streams and rivers, and the replenishing of groundwater.

Temperature increases will leave the atmosphere thirsty for water. Greater amounts of evaporated water are expected to be sucked up from dams and soils by the atmosphere and transpiration will increase in plants. The only crop that has been studied as part of South Africa's climate change modelling is maize which, in 1996, accounted for 71 per cent of grain harvested that year and 58 per cent of land under crop. Arguably South Africa's most important grain product, 60 per cent of maize grows in the drier western part of the country's maize region and 40 per cent in the wetter east[37]. The most abundant regions occur in the Free State and North West provinces, followed by the Mpumalanga Highveld.

The population of South Africa, Swaziland and Lesotho – where the staple food is maize – is expected to climb from the current estimate of around 45 million to about 70 to 90 million by 2035[38]. In the hotter, drier climate predicted by climate models, annual yields will drop between ten and twenty per cent by 2050, even though the region requires an annual three per cent increase to meet its future food requirements. The marginal western plantings are expected to suffer most, while the wetter eastern range of maize will be least affected. Various adaptation mechanisms have been suggested: farmers could change their crops to more drought-resistant species such as millet or sorghum, or they could shift from crops to livestock farming. South Africa certainly cannot hope to lean on its water supply to save its maize crops.

The one thing that characterises all famines – those which took place in the 1300s, and those which happen today in Africa – is that the poor carry the full weight of catastrophe. Whenever a seller's market dominates, consumer demand pushes up prices. When that market is food, it is always the poor who will go to bed hungry. Not those of us who are accustomed to foraging along bulging supermarket aisles. Nor those of us who enjoy the indulgent middle-class luxury of self-exploration as we backpack around different areas of the globe. The adventure-seeking student's momentary discomfort or concern about the next meal can easily be rectified or escaped. That palpable insecurity bears no relation to the abject poverty and suffering of Africa's marginal people or the world's poor. Those in the poverty trap have no escape route.

Dr Scott Drimie, senior research specialist with the Integrated Rural and Regional Development Unit of the Human Sciences Research Council (HSRC), believes that the South Africans most vulnerable to food insecurity are those still living in former Bantustan areas. Here they are often cramped up into marginal land, endure grinding poverty and are situated far from government services. Their children – whose little bellies are often bloated with kwashiorkor – are the ones most likely to succumb if famine strikes. For these people there is simply no margin for error.

Helping the poor to help themselves, as many do-gooders will, is not as easy or simple as giving each household a bag of seed. Planting crops requires capital – land, the means to plough, water, and time – and the investment is not always guaranteed to bring returns. If the rains fail, a season's worth of work is lost and there are no savings to fall back upon.

Many people in these areas would prefer wage labour and it is this imperative that drives urbanisation. About 55 per cent of South Africans now live in cities, creating a whole new sector of potential food insecurity that Drimie believes is not fully understood. Informal housing, on the periphery of the country's urban centres, provides meagre shelter for millions of people who have moved to the city in search of work. Many don't find work. Unemployed, they have no cash income to buy food. Newly urbanised, they also have no arable land to grow food. What will happen to these people living on the fringes of the economy if maize shortages of twenty per cent hit the country in the next few decades? If they don't have money to buy food and no means to grow their own, what do they do?

Institutionalised religions and centuries-worth of accumulated ego have given humans the belief that we have dominion over the planet, that we are somehow removed from its processes and are superior to the workings of nature. It's time for a reality check. The same rules governing population dynamics in any animal or plant species apply to us in equal measure. An ecological speed wobble will eventually develop when a species begins to overrun its habitat, using up the resources and outcompeting other creatures with which it must cohabit. So successful is *Homo sapiens* that it has exploited nearly every available niche across the planet, including some of the most inhospitable places on Earth. Humans have overrun the savannas and prairies,   slashed and burned into forests, paved over, built up, flown around and dug under in the process of exerting our precarious position as the dominant species on the planet.

Gods we are not. Our fallibility is made manifest when a mountainside disintegrates into a surge of mud, when a tidal wave lays waste a string of seaside villages, when the rains fail, or when our technologies collide in a spectacular crush of metal and glass. Carl Jung stated that the only real danger that exists is man himself: 'We know far too little [of man]. His psyche should be studied because we are the origin of all coming evil[39].'

Within the next fifty years the global population is expected to reach eight to ten billion[40]. Optimists say we will continue to keep the banquet table groaning under the weight of ample food while our inherent good will keep spreading the abundance. Pessimists say that soils are faltering, the struggle for water will bring out the war-monger in us, and there's only so much that our species can extract from this spinning piece of rock with its miniscule shroud of life-supporting biosphere. When the fight for life comes, the animal will start to eat itself.

# Chapter 14

Fouling the atmosphere is the first truly globalised form of pollution
Photograph © David Larson/The Media Bank/africanpictures.net

# Six degrees
# of separation

*Once the last tree is cut and the last river poisoned, you will find that you cannot eat your money.*

Anonymous

I started this journey in search of a smoking gun in the riddle of climate change. The question I took with me as I boarded the *SA Agulhas* at the very start of it all was this: does it matter? Why does it matter that a shoebox-sized island in the Southern Ocean, which few people will ever even see, dries out and is over-run by alien species? The same question went with me into the Knersvlakte in the searing heat and sat in my mind while I swatted mosquitoes next to a sizzling braai fire in the Kruger National Park. It simmered away when I watched a scientist perform ultrasound on a tortoise in the kitchen of an obscure little bed-and-breakfast in the rural Cape. No doubt it was somewhere in the recesses of my mind when those elephants charged. It was there when I lay in bed listening to nocturnal frogs welcoming in the winter rains with song. It followed me to my desk the next morning where I ate marmalade toast, waiting for the computer to fire up for another day on the job.

It's a really good question. Why, indeed, does it all matter? Any notion of good or bad is a human construct. If you are of the view, as I am, that humans are part of the ecosystem, then the changes we are bringing to the planet are natural and part of the same background noise that has rumbled from this great engine called life since it all began.

We live in a system that is in constant flux. Climate has changed more times than you can count on both hands. Species have moved and colonised since the dawn of life. Humans have simply accelerated the process. The bad news is that this could result in our own extinction as well as that of many of the species co-existing with us. But even that is part of the natural ticking over of things. Every species dies out eventually.

So why does it stick in the craw? An economist pointed out to me that, while some arguments may be theoretically sound, they are not necessarily ethical. Take the case of Lawrence Summers, Harvard-educated economist who served as chief economist at the World Bank from 1991 to 1993. He circulated an internal memo in '91 in which he suggested a solution for the problem of the accumulating toxic waste produced by rich countries. Simply pack it up and send it off to undeveloped countries for disposal along with the kind of monetary incentive that no cash-strapped or greedy administration can refuse. The economic logic was this: the cost of health-damaging waste is measured in terms of lost earnings owing to disease and death. In countries where wages are lowest, the loss of earnings will be lowest, so it makes sense to dispose of toxic waste in those countries.

Summers pointed out that concern over poisons which produce prostate cancer, for instance, is much higher in a country where the population is likely to live long enough to develop prostate cancer. In poor countries, where fewer people live to old age, no one really worries about exposure to carcinogens if they are unlikely to live long enough to see the cancers manifest. So why not let the rich industrial world pay developing nations to take its toxic rubbish? In that way, poor nations could benefit from a rich nation's problem. It's win-win. 'The economic logic behind dumping a load of toxic waste in the lowest wage country is impeccable,' Summers said in his memo.

The Brazilian Secretary of the Environment Jose Lutzenberger (his German

extraction explains the last name) replied, stating that the logic was insane and
evidence of:

> Unbelievable alienation, reductionist thinking, social ruthlessness and the
> arrogant ignorance of many conventional 'economists' concerning the nature
> of the world we live in.

Lutzenberger was fired by his government. Summers went on to complete his term
before joining the US Treasury Department and later became president of his *alma
mater*. No further comment.

Since we are talking about value systems, let's take a moment to consider them
in terms of rands and cents. The law of supply and demand in a free market economy
dictates that when demand outstrips supply, the price soars. When demand drops,
so does the financial value of something. A few thousand years into human trade
history, we have a curious situation where a supermodel can strut for a daily fee of
no less than £10 000, a Van Gogh painting auctions for nearly £30 million, and
soccer-player David Beckham changes hands between soccer clubs Manchester
United and Real Madrid for £25 million. The way of the free market dictates that
a pair of designer Italian shoes can fetch $500 in a New York boutique and a big
name Hollywood star can ask for $20 million per movie. It's a simple balance of
power: many buyers and a perceived desirability inflate the value of the product
and send the seller laughing en route to the bank.

We place different values on different things. In terms of basic human value, a
bricklayer is equal to a medical doctor: a murderer would receive the same jail
sentence for killing either of them. But the doctor's knowledge and skill come at a
higher price than those of the bricklayer, which means the doctor's family could
sue the murderer for a great deal more money than that of the bricklayer. This is
how South African Airways justified a R242 million golden handshake to dismissed
CEO Coleman Andrews. Meanwhile, the minimum wage of a fulltime domestic
worker in urban South Africa is pegged at R930.15 per month.

Then there is the value system that treasures human heritage and our own cleverness – ancient Greek sculpture, the Archimedes screw and the Egyptian Book of the Dead. Gunpowder, chopsticks, ceramics and horticulture. The loom, printing press, steam engine and light bulb. Airplanes, nylon, microchips and nanotechnology: milestones in the development from rudimentary cave dwellers to explorers of space. We mark our civilisation with the benchmark of masterful art and sophisticated thinking. Libraries are repositories of knowledge and art galleries are crammed with incomprehensible wealth in the currency of colour on canvas, stretched neatly into frames and discreetly lit for our viewing pleasure. These are the price tags of wealthy nations.

In another world where a billion people live on less than one dollar a day[1], the priorities become a little more fundamental: how will the single mother of four, who owns one ox, provide for her hungry children *today* after another season of failed rains? The ox is her only security – it can till the soil on her plot and bring in a small income from working her neighbours' land. But the woman's children are hungry *now*. She can sell the ox for, say, R3 000 to stave off the hunger pains immediately but she will then forfeit tomorrow's full bellies.

What, then, is the value system of 6.4 billion people, half of whom live in cities[2]? How do they prioritise the importance of things when many are so far removed from natural systems that they have forgotten the link between boiling the kettle and harnassing the energy needed to make that cup of tea?

In 1997 a group of scientists and economists put their heads together and connected the dots. They calculated the value of goods and services provided by nature to humanity: fibre for clothing; timber for furniture and housing; fodder and crops; regulation of the climate and keeping the atmosphere breathable; capturing and purifying fresh water; energy; nutrient recycling (including managing our daily bowel movement); drugs that keep us healthy . . . You name it – if you eat, drink, sleep or take a headache tablet, you somehow need nature to provide a service to you.

The group's final sum in their calculation amounted to $33 trillion or more annually – twice the gross national product of all the countries in the world for

1997[3]. Their argument, I have since discovered, is fraught with problems. Bio-economics is a fairly recent school of thought and scholars have still not reached consensus on how to define nature and which aspects should be given a price tag and which should not. Nevertheless, to the question of how much nature is worth, one critic of the $33 trillion figure was content to settle for this: it's worth a lot[4]! There may be problems in the fine print of how the exact figure was reached but the overall concept is sound and it certainly started people talking.

In the global balance sheet of natural resources, we are using more than we have in the bank. We are living on borrowed time. And here I am not quoting the ranting of eco-fascists or doomsday fanatics; these words come directly from the United Nations and the World Bank[5, 6]. Let's not get bogged down in numbers – by how many tons our fish resources have reduced; the extensive hectares deforested and turned into desert; the vast quantities of poisonous lead burned into the air; the countless number of square kilometres of ground sterilised by irrigation or encrusted beneath tar and cement. I would prefer to talk about how we ascribe value to aspects of nature that are not traded on the Johannesburg Stock Exchange (JSE) or Wall Street – things to which we cannot easily attach a price tag.

Consider this – people who own dogs are found to be six times more likely to survive a heart attack[7], probably because owning a dog stimulates your nurturing instinct and the daily walks do wonders for your cardiovascular health. You can thank the wolf family for that benefit. If you find yourself behind bars, your stress levels will be significantly lower if you have a view of nature or farmland from your cell window. And next time you go for root canal treatment, ask your dentist to install plants in the surgery and pictures of nature on the ceiling because they will help to calm your nerves[8].

Edward Wilson, who penned the facts noted above in *The Future of Life*, lists the approaches we can adopt with regard to valuing nature. Anthropocentrism, he says, dictates that 'nothing matters except that which affects humans'. The problem with taking an economic approach to nature is that it is inherently anthropocentric. Economics is all about goods and services. As soon as we use that standard to measure the value of nature, the outcome is limited to those aspects of nature that benefit or support our survival and comfort. First, these aspects exclude a great deal of nature. Second, we don't understand ecosystem functioning well enough to be able to make a solid judgement call. A food plant is easily measured in rands and cents, but what about its pollinating insect that we have not yet discovered?

Wilson offers other approaches. Pathocentrism, he says, attributes greater value to 'chimpanzees, dogs, and other intelligent animals'. Extrapolate this to a scenario that values all intelligent people but fails to value the dim-witted, then you're knee deep in an ethical quagmire. Meanwhile, biocentrism acknowledges the intrinsic right to life of all living organisms[9]. It might assist, in view of this, to remember that we are related to every other living organism on the planet. We share a common ancestor, a single-celled organism that lived three and a half billion years ago and looked something like present-day bacteria[10].

In terms of these value systems, does it matter that the extinction rate is a thousand times higher now than in fossil records, or that in the future it is expected to be ten times higher than it is at present[11]? In the cold, hard light of geological time, it does not matter at all. About 99 per cent of all the life that has ever existed on the planet has fallen by the wayside. Sometimes it has been through a slow process of natural attrition; at other times it has been fast and cataclysmic. But always natural. This is probably the first time that this level of extinction is being achieved by the success of a single species out-competing others. Certainly it's the first time it is being done knowingly, surely the biggest irony of all.

We could try to persuade the planet's governments and decision-makers to act for the conservation of the environment by taking a bio-economic approach to calculating the exact monetary value of everything. We could point out, for instance, that the Cape Floral Kingdom brings in around R10 billion annually to the Western Cape coffers[12]. That figure was arrived at after calculating the annual income from ecotourism (R6 406 million), the harvesting of fynbos for thatching, cut flowers and the like (R78 million), and exploiting marine products along the region's coastline (R1 300 million). The Cape honey bee pollinates R1.8 billion worth of agriculture in the Western Cape every year, including berries, fruit, seeds, nuts and vegetables. Over and above that, beekeepers make another R8.64 million per year hiring out hives to farmers, while a further R13.9 million comes from fynbos honey. While 80 per cent of the province's hives depend on eucalyptus trees for honey production (which are 'alien' to this location, having been brought here from Australia), they are entirely dependent on fynbos for foraging throughout the winter months. Without fynbos, Western Cape agriculture would be seriously compromised. Suddenly the need for its conservation takes on a whole new importance, does it not? I doubt, too, that anyone has calculated the value of the increased productivity and saved medical expenses of all those people who enrich their mental and physical wellbeing by hiking the fynbos-covered mountains of the Cape.

These figures may be contentious but, for the sake of argument, let's use them as a rough estimate to prompt us to think about value next time we lose ourselves in a thousand-yard stare across a natural landscape. Wilson points out that it would be impossible to replace the services with which nature provides humanity. But, if this is not sufficient to persuade decision-makers to act, perhaps this will be: our own extinction could be just beyond the horizon.

Five mass extinction events have hit the planet since the first single-celled organisms split and multiplied. These events have snuffed out more life than we can imagine in the relative blink of a *Brontosaurus's* eye. The first happened about 440 million years ago, towards the end of the Ordovician period. Next came the Devonian mass extinction at about 365 million years ago. Both put paid to 80 to 85 per cent of life on Earth[13]. Biodiversity bounced back, only to be hit by the third great extinction during the Triassic period about 210 million years ago where 70 to 75 per cent of species threw in the towel. Then, about 65 million years ago, during the most famous extinction of the Cretaceous, dinosaurs were hurried off this mortal coil by a massive asteroid collision. But the humdinger of all extinctions, the undisputed heavyweight obliterator of life, came at the tail end of the Permian period about 250 million years ago. A catastrophe of such magnitude struck the planet that 95 per cent of all life disappeared, leaving a fossil record almost wiped clean of species.

Michael Benton, Professor of Vertebrate Palaeontology and Head of the Department of Earth Sciences at the University of Bristol, documents the various theories behind this hyper-extinction in his book *When Life Nearly Died*. In his opinion, the culprit was runaway global warming triggered by volcanic eruptions.

It probably happened like this: massive volcanic activity in Siberia pumped the atmosphere with enough carbon dioxide to trigger a relatively minor global warming event, just enough to start melting polar icecaps. The melting caused a release into the atmosphere of a compound called methane hydrate, which was frozen into polar sediments. This form of methane, a highly potent greenhouse gas, tipped the system into overdrive, creating runaway global warming. Increased heat, along with acid rain caused by the sulphur dioxide ($SO_2$) also released during the volcanic activity, killed plant life across the planet. Once the global life-support system collapsed, it took the rest of the food chain down with it. Herbivores starved to death as their grazing wilted. Predators followed suit as their own food source rotted around them. Life at the bottom of the ocean suffocated as ocean currents stopped churning, depriving the seabed of oxygenated water[14].

Calculating the total temperature increase that almost finished life on Permian Earth, Benton puts the figure at 6°C. The Intergovernmental Panel on Climate Change (IPCC) calculated its own upper estimate of temperature increase in the next one hundred years: 5.8°C[15].

But more recent work suggests these increases could be much higher. A meeting of scientists in Berlin in 2003 revised the temperature increase during the next century at anything from 7°C to 10°C[16]. Following this, Oxford University's climateprediction.net project ran more recent and extensive climate change models. An estimated atmospheric carbon dioxide level of 550 parts per million (ppm) – expected to be reached by mid-century – could bring about a maximum increase in some parts of nearly double the IPCC's upper figure: 11°C by 2100[17]. Some scientists criticise this figure, but it is part of the slow refinement of the modelling process so it should not be dismissed outright.

The researchers pointed out that it is impossible to define the 'safe' upper limit of atmospheric $CO_2$[18]. At the same time, the British MP assigned to the International Climate Change Taskforce, Stephen Byers, told the BBC that a 'carbon dioxide concentration of over 400 ppm would be dangerous'.

We are already committed to a 2°C increase in terms of how the planet's thermostat distributes heat. A lag exists in the system because it takes a long time for the heat trapped by greenhouse gases to be distributed throughout the system.

Changes in how the climate system performs will not be smooth or gradual either, but will tend to lurch[19], producing sudden changes. It's something like putting an old whistling kettle over a flame – more and more energy is added to the contents until suddenly the kettle erupts in a screech.

'Just as the slowly increasing pressure of a finger eventually flips a switch and turns on a light, the slow effects of . . . changing atmospheric composition may "switch" the climate to a new state,' says veteran environmental writer James Clarke, quoting a US National Academy of Sciences report in his book *Coming Back to Earth*[20]. The lag in the system means the consequences of global warming that are being witnessed today – large-scale glacial retreat; the earlier arrival of spring; plants and animals shifting their home ranges; and massive coral bleaching – are the result of greenhouse gas emissions put out during the 1960s and 1970s. Emissions from the current decade will only play their hand after about three to five decades. Even if we were to shut down factories across the globe, to ban every single car, to reduce rice paddies and mass cattle feedlots, and to cut greenhouse gas emissions to pre-industrial levels in one day – which is only good as an intellectual exercise because it would be impossible to do practically, it would take decades, if not centuries, for the momentum to slow down.

The WWF released a statement at about the same time as the Oxford University findings in which it said the next two decades could see a threshold reached that would tip the scale dangerously, triggering disruption to the Earth's climate which could result in an even greater occurrence of rising sea levels, storms, droughts and larger-scale extinctions than already experienced across the globe. The IPCC's notoriously conservative chairman, Dr Rajendra Pachauri, also said that dangerous levels of $CO_2$ had already been reached, in spite of the climate change watchdog's attempts to encourage the reduction of greenhouse gas output. Speaking at a conference of Small Island Developing States held on the island of Mauritius in January 2005, Pachauri told representatives of 114 countries that the 'window of opportunity' for cutting emissions and stalling the dangerous decline into runaway global warming was 'closing fast'[21].

In 1992, representatives of over 120 countries gathered in Rio de Janeiro where they acknowledged that the climatic system is a shared global resource. By signing the United Nations Framework Convention on Climate Change (UNFCCC), they committed themselves to taking action against the problem of global warming and greenhouse gas emissions. Another gathering took place in Japan in 1997 where the Kyoto Protocol came into being. This global agreement set targets for emissions reduction and timeframes in which it should be done. Kyoto finally came into effect in February 2005. The agreement requires signatories to cut back on the amount of emissions produced during a set target period, the first being 2008 to 2012, as well as to encourage clean development in non-industrial countries.

Kyoto's biggest problem was to create some kind of equity in its emissions reduction scheme. The huge polluters in the north needed to be encouraged to cut back but the system could not handicap non-industrial countries where development is needed to improve the standard of living of their populations. Kyoto's solution builds cross-subsidisation into the system. Developed countries, which already have costly gas-emitting infrastructures driving their economies, will find it difficult to cut back. Developing countries, which have yet to build up their infrastructure to the same level, will find it easier to change their potential emissions output by switching their development plans to sustainable and renewable technology when they begin to put that infrastructure in place. Kyoto structured itself to use those countries that cannot cut back easily to incentivise and fund those countries which can.

An example would work something like this. A developer in South Africa plans to build low-cost housing with solar-heated water units. Norway funds the solar units.

Norway meets its Kyoto obligations, South Africa receives 'aid' for development, and the environment benefits by not having more greenhouse gases generated to heat water for those new houses. The currency for the transaction is the carbon credit – South Africa has carbon credits for sale in terms of solar units and Norway buys carbon credits by paying for them in cash with the principles of Kyoto to mediate. Under Kyoto, industrialised nations – so-called Annex I (one) nations – pay for non-industrialised nations – or non-Annex I (one) nations – to develop cleanly.

The problem is that this is a straight-line solution in a world that's not filled with straight lines. The United States and China are the two biggest greenhouse gases emitters. The US is an Annex I country and China a non-Annex I country. Given the global trade pressures and historical tension between the two countries, the US is unlikely to find the political will to pay for China to clean up its act. Consequently, Kyoto has been described as politically untenable for the Republican administration.

The reason this matters is because the US, with only five per cent of the world's population, produces twenty per cent of the greenhouse gas emissions. Several paces behind is China with 14.74 per cent of emissions, followed by the European Union with 14.05 per cent and the Russian Federation with 5.72 per cent[22]. India comes in fifth place with 5.48 per cent and Japan with 3.98 per cent. The figures drop off sharply from there. For the planet to survive, emissions need to be reduced quickly and drastically. As the biggest emitter, it is essential that the US puts its shoulder to this particular plough. In 1997 the US signed Kyoto under Democratic President Bill Clinton. But when his Republican counterpart moved into the Oval Office for the next presidential term, George Bush shied away from ratifying the agreement.

The balance of power was not always like this. The mass burning of fossil fuels started in the mid-1700s, but it was mostly coal that pulled side by side with wood in the harness of industrialisation. It was only towards the 1850s that commercial exploitation of oil wells began. By 1910, oil still only accounted for five per cent of the world's energy supply. It took a boom in the vehicle industry in the 1930s before oil streaked ahead of wood and coal in the energy race[23]. Soon the industrial world's demand for oil reshaped global geopolitics. This heralded the rise of the hydrocarbon economy and, by the end of the twentieth century, 90 per cent of energy used globally came from fossil fuels[24].

Developed countries since became increasingly dependent on cheap oil to fuel their heavily industrialised economies. Continued growth and the comfortable lifestyles of First World electorates were dependent on their country's ability to find cheap oil. *The End of Oil* author Paul Roberts says this scenario played itself out under an entirely new set of global rules with government foreign policy and oil companies manipulating global oil markets:

> In a remarkably short time, oil had moved to the very epicentre of geopolitics. Just as nineteenth-century imperial powers had competed for the colonies with the best sugar and tea and slaves, the industrial powers of the twentieth century manoeuvred for the choicest oil regions. Driven by the ravenous demand for oil, Western governments and their able assistants, the international oil companies, vied for control over the hapless oil states of Venezuela, Mexico, Sumatra, Borneo, and especially the Middle East, where European and US diplomats redrew the map to maximise access to oil. As one French diplomat declared during a period of particularly frenzied boundary drawing, 'He who owns the oil will own the world'.[25]

As dependent on oil as a junkie is on crack, the US demand for oil has skewed its game in the field of geopolitics, a bias that has resulted in two US incursions into Iraq. The first Gulf War was entirely about oil. Kuwait, fearing the strength of Iraq in the region as well as President Saddam Hussein's attempts to dominate the market, flooded  the market with oil. The price of crude oil plummeted, which Hussein read 'as tantamount to economic war' and invaded the country. US President George WH Bush responded by dispatching troops into the region in January 1991.

A decade later, supposedly to halt global terror, dethrone a tyrant and deliver democracy to the region, the next George Bush set his troops on Bagdhad with the Britons hurrying to their assistance. Saddam Hussein had purportedly been cooking up weapons of mass destruction (WMDs) and was aligned with the al-Qaeda terrorist group.

'My fellow citizens,' said Bush in his speech to Americans in March 2003, 'at this hour, American and coalition forces are in the early stages of military operations to disarm Iraq, to free its people and to defend the world from grave danger.'

Roberts explains that while Hussein was by no means a gentlemanly or humane leader, beneath this thin veneer of American philanthropy lay the true reason for the attack: US oil imperialism needed to be advanced in a market where prices were out of control owing to the tinkering of the puppet master – the Organization of Petroleum Exporting Countries (OPEC). So misinformed was the US public by now from the constant spin by the White House (and, in many cases, the complicity of a jingoist media) that three quarters of the American public believed that Hussein was responsible for the September 11 attacks on the World Trade Centre in Manhattan[26]. Not surprisingly, WMDs were not found and United Nations Secretary General Kofi Annan labelled it an illegal war. Yet the damage has been done and, as these words are written, Iraq is more unstable now than before.

> 'American oil lust is a mixed blessing,' states Roberts, '. . . such heavy dependence on foreign oil makes the United States vulnerable to disruptions in supply and to energy "blackmail" and has, in addition, fostered a long tradition of doing whatever is necessary, covertly or overtly, to ensure that the United States – and US oil companies – have access to world oil supplies.'

It is no secret that the Bush family roots sink deep into the oil industry. They go back a hundred years to George junior's grandfather, Prescott Bush, who cut his teeth in the international oil business during a two decade-long term as the director of the arms manufacturer and oil-services company Dresser Industries. George HW Bush worked for Dresser and ran his own offshore oil-drilling business, Zapata Offshore. George W Bush mostly raised money from investors for failed oil operations[27]. Since the 1980s, the Bush family was well connected with Enron, the giant US energy company. George HW Bush gave key personnel sufficient clout in Washington to later allow George junior, then on his presidential campaign trail, to head straight to Enron for contributions. This made Enron one of the most powerful corporates in memory.

Dick Cheney resigned as CEO of the oil services company Halliburton to join George W Bush in the White House. He would later be accused of securing lucrative business for Halliburton while serving his term, including facilitating a no-bid contract to rebuild Iraq. Condaleezza Rice, former national security adviser and now US secretary of state, served on the board of directors of the Chevron Corporation – the massive California-based energy company – for a decade before her appointment by Bush. Rice's connection to the company hit the headlines when Chevron named one of its oil tankers after her in 2001. Writing for the liberal online magazine Salon.com, Damien Cave said 'no administration has ever been more in bed with the energy industry . . . the Bush administration's ties to oil and gas are as deep as an offshore well'[28].

The US started out as the world's largest producer of oil and other energy. Now it is the largest consumer. One out of every four barrels of oil produced in the world is burned in America, asserts Roberts, and this 'enormous, apparently limitless appetite exerts a ceaseless pull on the rest of the world's oil players and on the shape of the world political order'[29]. You can see the conflict of interest: what CEO is going to make decisions that will shrink the board's pay package or risk upsetting the shareholders? For the United States to reach the goals set by Kyoto and perform its crucial part in slowing anthropogenic global warming, the country would need to reduce its emissions by 70 per cent by the end of this century. This means that not only would the current American population have to cut back radically on their ample lifestyles, but so too would the future population – which is expected to become larger and increasingly wealthy.

The White House ethos is to deny the complicity of humanity in climate change. It was best expressed by Oklahoma Senator James Inhofe, chairman of the Senate

Environment and Public Works Committee, who said that global warming is 'the greatest hoax ever perpetuated on the American people'. Pollution and its consequences – which are so intrinsically linked to how people live their lives and hence to their own complicity – also seem to 'strike a raw nerve'. This debate has become a political milieu that took a most sinister turn in 1998 when a group of scientists published a paper in *Nature*. The paper argued that of all the factors changing temperature globally during the past 600 years, greenhouse gases were the main driver behind temperature increases in the 20th century[30]. A political backlash followed early in 2005 when the Chairman of the Committee on Energy and Commerce with the US House of Representatives Joe Barton, an ardent climate change sceptic, requested information on the scientists. *The Washington Post* described the action as an intrusive and far-reaching witch hunt, with a possible 'chilling effect on the willingness of people to work in areas that are politically relevant'.

A lack of consensus in the scientific community is further complicating things. While most scientists agree about the general causes and effects of climate change, it's a sufficiently complex science for there to be some disagreement about the finer points. There are also some scientists who question whether the prognosis is as bad as the IPCC says. The uncertainty is enough for the industrial giant to stall on committing to emission reductions and the economic consequences that will result. The US government has been accused of 'cherry picking science to suit policy agenda and suppressing science that it did not like'[31].

Some question whether Kyoto will even work. Another theory is that the US refuses to sign Kyoto because, if industry is curtailed in the US, factories will simply move abroad to poor countries where human rights and pollution controls are not as stringent. This is almost certainly not the case, although the kind of thinking in Lawrence Summer's memo, which actively encouraged sending pollution to the poorer nations, probably still exists in the upper echelons of US government today.

I'm taking a hard line on this because the US government realises there is a problem and yet fails to act to bring about change. Early in 2004 a secret report produced by the Pentagon and suppressed by the US Department of Defence was uncovered. In it the Pentagon warned the US president that climate change is a far greater threat than global terrorism. The report said that:

> Abrupt climate change could bring the planet to the edge of anarchy as countries develop a nuclear threat to defend and secure dwindling food, water and energy supplies. Disruption and conflict will be endemic features of life. Once again, warfare would define human life.[32]

Across the Atlantic, British prime minister Tony Blair has gone on record saying that global warming is 'the single most important issue we face as a global community'[33]. Yet the United States remains the only G8 member which has not backed the Kyoto Protocol. UK environment secretary Margaret Beckett called on the US in May 2005 to accept the 'overwhelming scientific evidence that climate change is happening and manmade'. She says:

> We think the science is incontrovertible. We want that understanding to be widened and deepened. Those who argue against the science are fewer and fewer in number and treated as less and less respected.[34]

But now let's turn to the great sleeping giant that will probably be tomorrow's huge problem. Forget Newcastle; in the world of fossil fuels, all coal-related idioms belong to China. Coal is by far the filthiest fossil fuel and China is the biggest producer and user of it. The country is a rising behemoth that hosts twenty per cent of the world's population (1.3 billion versus the 290-odd million in the US), with the third-highest gross domestic product, and growing at three times the rate of the global average[35]. But it needs coal to keep the engines of economic growth firing. China is the second-largest producer of electricity globally. Nearly 80 per cent of that electricity comes from chucking coal down the chutes of power stations across its massive girth[36]. The country's government has made its policy of continued reliance on coal well known so, as the demand for power doubles by 2020, the increase in annual coal consumption will climb to about 3.5 billion tons.

China's contribution to air pollution across Asia is massive. Airborne ash and soot – black carbon from the burning of coal and soot particles from wildfires and domestic wood use – are clogging up the air and altering regional climate. For the past twenty years the southern part of China has seen an increase in flooding while the north is becoming drier and drier. While scientists admit that they cannot tell *exactly* how much the air pollution is responsible, it most definitely has played a part. The particulate matter causes atmospheric warming but keeps the ground cool. It also causes smaller droplets to accumulate during the formation of clouds, which ultimately reduces the amount of rain that falls[37]. Nitrous dioxide, emitted through the combustion of fossil fuels and biomass, has accumulated in the troposphere above China by 50 per cent between 1996 and 2004 (while a large reduction has been noted over parts of Europe and the US)[38]. Three out of four Chinese urbanites live below the national air quality standard and the region is the most severely affected by acid rain globally[39].

Writing in *Nature*, Peter Aldhous comments that China has the potential to single-handedly emit enough $CO_2$ to negate all other nations' efforts to control their greenhouse gas emissions. The future of China on the world emissions stage is dependent on its ability to develop clean coal-burning technology. Ironically, China's size (in land and numbers) is going to impact on the rest of the world, but it is global trade and foreign investment that feed into the growth rate[40].

The US and China aside, the rest of the industrialised and developing worlds should remember their own culpability. This includes the wealthier sectors in countries such as our own with the so-called two economies where rich and poor are juxtaposed. South Africa produces about 1.23 per cent of global greenhouse gases[41] but per capita it is among the worst emitters of greenhouse gases in the world. The World Resources Institute ranks our country as the nineteenth biggest emitter in the world and our per capita output is more than 3.5 times the average for the developing world[42]. Comparing per capita production of greenhouse gases, measured in equivalent tons of carbon dioxide ($CO_2$), the US produces 23.8 tons of $CO_2$ per person. South Africa produces 9.16 tons per person. China only produces four tons per person. Australia, although only responsible for 1.46 per cent of annual greenhouse gas emissions, produces 25.3 tons of $CO_2$ per person. That's higher than the US. As an Annex I country, Australia has also avoided ratifying the Kyoto Protocol.

The bad news is that you and I are probably not far behind the big consumers and emitters. If you purchased this book, then – like me – you are probably in the socio-economic bracket that brings you closer to the high-energy lifestyle of an urbanite in the US or Australia than to that of the subsistence farmer in the rural Eastern Cape. South Africa's middle class per capita production of $CO_2$ is probably much higher than the national average, which is reduced by the vast majority of the country's poor whose energy demands and $CO_2$ output are constrained by poverty.

The picture looks like this in table form[43]:

| Country | Status | Kyoto ratification | % of global emission | Per capita emissions, measured in tons of $CO_2$ |
|---|---|---|---|---|
| USA | Annex I | No | 20.64 | 23.8 |
| China | Non-Annex I | Yes | 14.74 | 4.03 |
| European Union | Annex I | Yes | 14.05 | 10.26 |
| Russia | Annex I | Yes | 5.72 | 13.2 |
| India | Non-Annex I | Yes | 5.48 | 1.83 |
| Japan | Annex I | Yes | 3.98 | 10.63 |
| Australia | Annex I | No | 1.46 | 25.3 |
| South Africa | Non-Annex I | Yes | 1.23 | 9.16 |

While not emitting large volumes of greenhouse gases, there are other countries that rank even higher than the US in their per capita output of carbon dioxide.

Kuwait produces 30.43 tons of $CO_2$ per person. The United Arab Emirates comes in at 35.93 tons and Qatar, the world leader in its per capita emissions, sends out 66.36 tons of $CO_2$ per person.

South Africa is responsible for roughly half of Africa's total greenhouse gas emissions with our energy-intensive economy and the cheap, dirty coal firing our power stations. Cheap electricity has given the country a competitive edge in its market presence. We produce more than half of Africa's electricity – 90 per cent of which comes from burning coal – and use more than half of the electricity ourselves[44]. Energy-intensive mining, heavy industry and manufacturing account for 45 per cent of our energy demands. Transport comes in next at twenty per cent of total energy consumption. Residential demands consume ten per cent, while commerce and agriculture both use three per cent[45]. Most of the resulting greenhouse gas emissions come from the energy sector, with significantly lesser amounts deriving from industrial processes, agricultural practices and waste.

As a non-Annex I country, South Africa is not required under Kyoto to reduce its current emissions during the 2008 to 2012 target period. The government has nevertheless indicated that it is aware of South Africa's considerable contribution to greenhouse gas emissions and has indicated its intention to clean things up. But this is mostly still only at policy level. Whatever happens from here, a ship this size is slow to turn.

Industry and transport are our two big offenders. Since coal is the main source of our greenhouse gas emissions, cleaner ways of burning coal and cleaner fuels *must* be found. Eskom and Sasol account for nearly 90 per cent of all coal burned in South Africa (at 65 per cent and 24 per cent respectively[46]). Industry needs to be more energy efficient. The government needs to impose strict emissions audits so that we can know who the polluters are and then keep them accountable.

As for transport, buses, mini-bus taxis, trucks and private cars are the bulk of our road users. Their exhausts spew out greenhouse gases, carcinogens and other noxious pollution daily as they clog up South Africa's roads. I'm lucky enough to work from home most of the time, but occasionally I have the misfortune of tackling rush hour on the byways of the peninsula, which are unbearably congested. Each year the same 20 km trip seems to take that bit longer and it's only getting worse as Somerset West and the northern suburbs bulge with ongoing urban sprawl. Of course, Johannesburg's traffic problems make those of Cape Town look like a kid's party.

It's a constant wonder that this problem has not been fixed. First there is the stress, loss of productivity and wasted time as hundreds of thousands of people spend an hour or two stuck in traffic twice a day, every day of the working week. (I've always maintained that life is too damn short to spend that much of it sitting in traffic.) Then there is the sizeable matter of road upkeep to handle the capacity, as well as the effects of the vast amounts of pollution puffing up from the grim, sluggish snake of traffic.

Government has set targets for reducing lead from petrol and sulphur from diesel[47] but a great deal more needs to be done. Why do we still have streams of vehicles with single occupants sitting bumper to bumper, clogging up the roads? Private vehicle users need to change their attitude and behaviour. Toll tax for single occupant cars and lanes reserved for full vehicles will encourage people to start lift clubs and to use their vehicles more thoughtfully. Why are larger municipalities not doing this? Once fuel becomes prohibitively expensive, it's likely that we will see increasing numbers of people offering to share their cars with passengers. Rush hour traffic is a simple problem to solve, requiring a combination of political will, capital and a dash of entrepreneurialism. It's extraordinary that the human species could put a man on the moon nearly four decades ago but has still failed to solve the problem of traffic congestion.

Public transport needs to revolutionise. It has to become more than the left-over option for blue-collar workers and the urban poor who have not yet made it into the stratum of society's car owners. Consider what rush hour would be like if trains and shuttle buses replaced the bulk of those single-driver vehicles. For every full eighteen-seater bus running between the city centre and the suburbs, there are likely to be seventeen fewer cars on the road. Why are the trains that criss-cross the large urban centres not safer and more reliable?

Why are 4x4 vehicles in the city not being taxed heavily or banned from city use outright, except for permitted use? Keep these vehicles where they are needed – off-road. Such wasteful, gas-guzzling status symbols are not necessary to ferry the kids to and from school. And while we are on the subject of conspicuous consumption – is a V8 necessary for darting about town when a 1.1 litre engine will do?

Having said that, a local car magazine editor friend of mine pointed out my naïve and reductionist take on the 4x4 issue. Many modern 4x4s do not use more fuel than similarly priced two-wheel drive counterparts. If government were to tax them on that basis, then such a penalty would also have to apply to all vehicles that produce over a certain amount of carbon per kilometre travelled. A tax of this nature would hit the poorest vehicle users too, many of whom drive oil-burning rust buckets by virtue of their economic marginality. Any of us who drive old, non-catalytic converter-equipped vehicles are probably as guilty of emissions as the fat cats in their 4x4s and V8s which, on the whole, are more fuel-efficient and more green than many old cars. And, as he pointed out  – which is the bit that really burns – there are high-end luxury German cars that, at higher speeds, use the same amount of fuel as my nine-year-old Citi Golf. As alternative fuels and hybrid cars filter onto the market, vehicle manufacturers are morally obliged to make these affordable, rather than to price them up as luxury vehicles – only targeting the middle class who wish to brand themselves as green conscious.

As for you and me, what can we do as individuals? A senior scientist told me it was pointless to go out of my way to switch off unnecessary lights at home and do similar energy-saving exercises. It sounds a bit defeatist, but he's probably right. The tiny bit of electricity wasted by a single light bulb is meaningless compared with the massive emissions that come from industry, mines and transport every day. Nevertheless, we can do our bit: shower quickly instead of bath, set the geyser to 55°C, install energy-efficient appliances in our homes, and try moving to solar panels where possible. The government could consider tax subsidies for energy-saving initiatives in the home. We can make wise decisions about the vehicles we use – buy smaller, fuel-efficient cars, start lift clubs and walk to the local shop instead of drive. We can pressure our municipalities and local government to act decisively to provide public transport. We can demand that national government and industry be kept accountable to the public for their policies, activities and emissions.

I have my doubts about how much of a difference any of these actions will make. The US and China seem set to carry on business as usual. While the US continues to secure its oil resources, it is now looking to tap new fossil fuels stored in the sediments beneath the seabed in the Gulf of Mexico. Deposits of methane hydrate crystals, formed by combinations of water and gases under high tectonic pressure, can be harvested as a fuel and the methane burned. The by-product is $CO_2$. *The Guardian* reported that 'scientists reckon there could be more valuable carbon fuel stored in the vast methane hydrate deposits scattered under the world's seabed and Arctic permafrost than in all of the known reserves of coal, oil and gas put together'[48].

So, while most of the scientific community and many politicians believe we are already committed to dangerous levels of climate change, the US is aiming to have this new and potent source of greenhouse gas commercially available as a fuel within a decade.

The question I take out of this journey is the following: given that we are a species of such remarkable ingenuity and aptitude, how bad does the situation have to become before we apply ourselves to finding solutions to the problem of global warming rather than continuing business as usual? And, when we get to that point, will it be too late? If it is not already too late.

At some point the industrialised world will have to rein itself in. If everyone on the planet lived to the excesses enjoyed by the people of the United States with current technology, we would need four more Earths[49] to provide the resources. In the past, pollution has always been fairly localised. But globalisation has finally been embraced in the pollution market. It reminds me of a rather inelegant but potent adage I heard a few years ago: having a 'no smoking' section in a restaurant is like having a 'no peeing' section in a swimming pool. By virtue of the mobility of the air around us, spewing filth into the atmosphere becomes the first truly global form of pollution. The pollution may go up into the sky over Gauteng, London or New York but it rains down its consequences in many places that are utterly non-complicit. The entire pool is being contaminated.

The UN's Millennium Ecosystem Assessment points out that the:

> Negative impacts of climate change will fall disproportionately on the poorest parts of the world – for instance by exacerbating drought and reducing food production in the drier regions – but the build up of greenhouse gases has come overwhelmingly from richer populations as they consume more energy to fuel their higher living standards.[50]

The IPCC first put the highest likely temperature increase within the next century at 6°C – the same temperature increase that wiped out almost all of life 250 million years ago. It may even get hotter. Not much is separating us from the great hereafter. The planet's sixth great mass extinction is upon us with the current rate of species loss. The question is not whether it will worsen, but by how much. Earth's biodiversity took about 80 to 100 million years to return to its former level after the Permian disaster[51]. If life in its current form were to die out, the planet would again recover – but we will probably have written ourselves out of the script.

# Notes and references

## Introduction

### Notes

1   This is now up to 6.5 billion.

### References

Houghton, J.T., Ding, Y., Griggs, D.J., Noguer, M., Van der Linden, P.R., Dai, X., Maskell, K. and Johnson, C.A. (Eds). 2001. *Climate Change 2001: The Scientific Basis*. Contribution of Working Group 1 to the Third Assessment Report of the Intergovernmental Panel on Climate Change. Cambridge: Cambridge University Press.

Lomborg, B. 2001. *The Skeptical Environmentalist. Measuring the Real State of the World*. Cambridge: Cambridge University Press.

Weart, S.R. 2003. *The Discovery of Global Warming*. Cambridge, Massachusetts: Harvard University Press.

## Chapter 1

### Notes

1   Luhr, 2004: 441

2   Watson, 1984: 20. In fact, Watson states that for every one per cent increase in oxygen, the risk of forest fire increases by 70 per cent.

3   Watson, 1984: 21

4   Luhr, 2004: 445

5   Rijsdijk, 2005: 'About 250 million litres in a lifetime of 70 years; 70 x 365 x 24 x 60 x 12 (breaths/minute) x 0.6 litre (volume per breath) = 265 million; rounded off to a manageable 250 million'

6   Watson, 1984

7   Luhr, 2004: 24

8   Although if you follow the M-theory (in essence, Membrane Theory) debate, you will understand why physicists are saying that the singularity did not exist. Rather, they now argue that the Big Bang was the result of the collision of two parallel universes that existed *before* ours. Our universe was brought into being by that cataclysmic event. This would suggest that time did, in fact, exist before the Big Bang and that our universe is just one of an infinite number. But M-theory is so far out there that it hurts to think about it too hard.

9   Rijsdijk, 2005

10  Luhr, 2004: 442

11  Rijsdijk, 2005

12  Luhr, 2004: 46

13  Luhr, 2004: 46 and Watson, 1984: 14

14  Watson, 1984: 14

15  Luhr, 2004: 46

16  Watson, 1984: 15

17  Luhr, 2004: 27

18  Watson, 1984: 17

19  Luhr, 2004: 443

20  Mielke, 1989: 47

21  Watson, 1984: 20

22  Luhr, 2004: 446

23  Watson, 1984: 17

24  Watson, 1984: 22

25  Bryson, 2003: 225

26  Luhr, 2004: 171

27  Bryson, 2003: 372

28  Rijsdijk, 2005

29  Ruddiman, 2005: 49

30  Ruddiman, 2005: 49

31  Ruddiman, 2005: 48

32  McGuire, 2002

## References


Black, Richard. 2005. BBC News: Alarm at new climate warning. [Online]. Available: http://news.bbc.co.uk/1/hi/sci/tech/4210629.stm. [Accessed March 2005].

Brown, Paul. 2005. *Guardian Unlimited:* Hotter world may freeze Britain: fifty-fifty chance that warm Gulf Stream may be halted. [Online]. Available: http://www.guardian.co.uk/climatechange/story/0,12374,1403798,00.html. [Accessed March 2005].

Bryson, B. 2003. *A Short History of Nearly Everything*. London: Doubleday.

Clarke, J. 2002. *Coming Back to Earth: South Africa's Changing Environment*. Houghton: Jacana.

Glick, Daniel. 2004. The big thaw. *National Geographic*. 13–33, September.

Houghton, J.T., Ding, Y., Griggs, D.J., Noguer, M., Van der Linden, P.R., Dai, X., Maskell, K. and Johnson, C.A. (Eds). 2001. *Climate Change 2001: The Scientific Basis*. Contribution of Working Group 1 to the Third Assessment Report of the Intergovernmental Panel on Climate Change. Cambridge: Cambridge University Press.

Klapper, Bradley. 2005. *Guardian Unlimited*: Global warming may kill off polar bears in 20 years, says WWF. [Online]. Available: http://www.guardian.co.uk/life/news/story/0,12976,1402330,00.html. [Accessed February 2005].

Kring, D.A. and Durda, D.D. 2003. The day the world burned. *Scientific American*. December.

Luhr, J.F. (Ed). 2004. *Earth: The Definitive Visual Guide*. London: Dorling Kindersley.

Lynas, Mark. 2003. *Guardian Unlimited*: At the end of our weather. [Online]. Available: http://www.guardian.co.uk/climatechange/story/0,12374,1055966,00.html. [Accessed February 2005].

Mantaigne, Fen. 2004. No room to run. *National Geographic*. 34–55, September.

McGuire, B. 2002. *A Guide to the End of the World – Everything You Never Wanted to Know*. Oxford: Oxford University Press.

Mielke, H.W. 1989. *Patterns of Life: Biogeography of a Changing World*. Boston: Hyman.

National Botanical Institute. 2001. The heat is on: Impacts of climate change on plant diversity in South Africa. [Brochure].

National Snow and Ice Data Center (NSIDC). 2002. Larsen B Ice Shelf collapses in Antarctica. [Online]. Available: http://nsidc.org/iceshelves/larsenb2002. [Accessed February 2005].

Rijsdijk, C. 2005. Personal email correspondence with the author. [Digital copy in possession of author].

Ruddiman, W.F. 2005. How did humans first alter global climate? *Scientific American*: 46–53, March.

Ruddiman, W.F. and Thomson, J.S. 2001. The case for human causes of increased atmospheric $CH_4$ over the last 5 000 years. *Quaternary Science Reviews*. 20: 1769–1777.

Simmons, R.E., Barnard, P., Dean, W.R.J., Midgley, G.F., Thuiller, W. and Hughes, G. 2004. Climate change and birds: perspectives and prospects from southern Africa. *Ostrich*. 75(4): 295–308.

Townsend, Mark. 2002. *Guardian Online:* Sea set to claim England's coastal treasures. [Online]. Available: http://www.guardian.co.uk/climatechange/story/0,12374,864532,00.html. [Accessed March 2005].

Vidal, John and Brown, Paul. 2003. *Guardian Online:* How climate change affects all our lives. [Online]. Available: http://www.guardian.co.uk/climatechange/story/0,12374,1029960,00.html. [Accessed December 2004].

Watson, L. 1984. *Heaven's Breath – a Natural History of the Wind*. Great Britain: Hodder & Stoughton Ltd.

Wilson, E.O. 2003. *The Future of Life*. London: Abacus.

World Meteorological Organization. 2002. Statement on the status of the global climate in 2002. [Online]. Available: http://www.wmo.ch/index-en.html. [Accessed: October 2003].


# Chapter 2

## Notes

1  Kring & Durda, 2003: 72

2  Kring & Durda, 2003: 74

3  Luhr, 2004: 123

4  Wilson, 2003: 23

5   Houghton et al., 2001: 2–7

6   Ruddiman, 2005: 50

7   Ruddiman, 2005: 50

8   Ruddiman, 2003

9   Ruddiman, 2003

10  Houghton et al., 2001: 2

11  Houghton et al., 2001: 4

12  Glick, 2004: 14

13  Houghton et al., 2001: 4

14  Klapper, 2005

15  Glick, 2004: 25

16  Domack et al., 2005: 681–685

17  Luhr, 2004: 451

18  Allen & Lord, 2004

19  Patz et al., 2005: 310–317

20  Allen & Lord, 2004

21  Patz et al., 2005: 310–317

22  Patz et al., 2005: 310–317

23  'At the end of our weather', 2003

24  Emanuel, 2005

25  'Hotter world may freeze Britain', 2005

26  Bryden et al., 2005

27  Houghton et al., 2001: 75

28  Houghton et al., 2001: 12

29  Houghton et al., 2001: 71

30  Stainforth et al., 2005: 403–406

31  A few years ago the South African government commissioned a study on the potential impacts of climate change on the country. Scientists from a variety of disciplines pooled their knowledge of climate, plants, animals and water resources and put the data through a series of computer climate modelling programmes used by the IPCC. The findings were published in the *South African Country Study Report* in the late 1990s but only released by government in the modified *Initial National Communication under the United Nations Framework Convention on Climate Change* in early 2004. The findings were already outdated by the time they were officially made public. However, many of the authors of those original papers are currently updating their findings using more refined modelling programmes. Some of the original *Country Study Report* results will emerge throughout this book – some of the updated findings will be included here and there. Once again, while the figures tend to vary slightly, the general trends mostly remain constant.

32 Kiker, 2000

33 Colinvaux, P. 2005.

34 Midgley et al., 2005

35 Thomas et al, 2004: 145–148

36 Simmons et al., 2004: 296

37 Thomas et al., 2004: 145–148

38 Hill, et al: 2002

39 'How climate change affects', 2003

40 Emslie et al., 1998

41 Mantaigne, 2004: 40

42 Desanker et al., 2001: 489

43 Schneider et al., 2001: 84

44 'The heat is on', 2005

45 Tadross, M., Jack, C. & Hewitson, B. 2005. On RCM-based projections of change in southern African summer climate. *Geophysical Research Letters*, Vol. 32, L23713.

46 Hewitson & Crane, 2005

47 Midgley, 2005: 47

48 Erasmus et al., 2005: 679–693

49 Simmons et al., 2004: 296

## References

Allen, M.R. and Lord, R. 2004. The blame game. *Nature.* [Published online: 01 December 2004]. Available: |doi:10.1038/432551a.

Bryden, H.L, Longworth, H.R. and Cunningham, S.A. 2005. Slowing of the Atlantic meridional overturning circulation at 25°. *Nature.* Vol. 438: 685, 1 December.

Colinvaux, P. 2005. Coping with interesting times. *Nature.* Vol. 437: 479, 22 September. Colinvaux, P 2005.

Desanker, P. and Magadza, C. 2001. Africa. In: McCarthy, J.J., Canziani, O.F., Leary, N.A., Dokken, D.J. and White, K.S. (Eds). 2001. *Climate Change 2001: Impacts, Adaptation, and Vulnerability.* Contribution of Working Group II to the Third Assessment Report of the Intergovernmental Panel on Climate Change. Cambridge University Press: Cambridge.

Domack, E., Duran, D., Leventer, A., Ishman, S., Doane, S., McCallum, S., Amblas, D., Ring, J., Gilbert, R. and Prentice, M. 2005. Stability of the Larsen B ice shelf on the Antarctic Peninsula during the Holocene epoch. *Nature.* Vol. 432: 681–685.

Emanuel, K. 2005. Increasing destructiveness of tropical cyclones over the past 30 years. *Nature.* Vol. 436, 686–688, 4 August.

Emslie S.D., Fraser W., Smith R.C. and Walker W. 1998. Abandoned penguin colonies and environmental change in the Palmer Station area, Anvers Island, Antarctic Peninsula.

*Antarctic Science*. 10(3): 257–268, September.

Erasmus, B.F.N., Van Jaarsveld, A.S., Chown, S.L., Kshatriya, M. and Wessels, K.J. 2002. Vulnerability of South African animal taxa to climate change. *Global Change Biology*. 8, 679–693.

Hewitson, B. and Crane, R. 2005. Consensus between GCM climate change projections with empirical downscaling: Precipitation downscaling over South Africa. *International Journal of Climatology*.

Hill, J.K., Thomas, C.D., Fox, R., Telfer, M.G., Willis, S.G., Asher, J. and Huntley, B. 2002. Responses of butterflies to twentieth century climate warming: implications for future ranges. Proceedings of the Royal Society of London. *Series B – Biological Sciences*. 269(1505): 2163–2171, 22 October.

Kiker, G.A. (Compiling editor). 2000. Terrestrial Plants Biodiversity Sector Executive Summary. In: Kiker, G. Climate change impacts in Southern Africa. Report to the National Climate Change Committee, Department of Environmental Affairs and Tourism, Pretoria, South Africa.

Midgley, G.F., Chapman, R.A., Hewitson, B., Johnston, P., de Wit, M., Ziervogel, G., Mukheibir, P., van Niekerk, L., Tadross, M., van Wilgen, B.W., Kgope, B., Morant, P.D., Theron, A., Scholes, R.J. and Forsyth, G.G. 2005. *A Status Quo, Vulnerability and Adaptation Assessment of the Physical and Socio-economic Effects of Climate Change in the Western Cape*. Report to the Western Cape Government, Cape Town, South Africa. CSIR Report No. ENV-S-C 2005-073, Stellenbosch.

Midgley, G.F, Kgope, B., Motete, N., Musil, C., and Mantlana, B. 2005. Modelling and experimental approaches to quantifying responses to climate. Gauteng: Talk presented to the National Climate Change Conference, October.

Patz, J.A., Campbell-Lendrum, D., Holloway, T. and Foley, J.A. 2005. Impact of regional climate change on human health. *Nature*, Vol. 438, 310–317, 17 November.

Ruddiman, W. 2003. The anthropogenic greenhouse era began thousands of years ago. *Climate Change*, 61(3): 261–293, December.

Schneider, S. and Sarukhan, J. 2001. Overview of impacts, adaptation, and vulnerability to climate change. In: McCarthy, J.J., Canziani, O.F., Leary, N.A., Dokken, D.J. and White, K.S. (Eds). 2001. *Climate Change 2001: Impacts, Adaptation, and Vulnerability*. Contribution of Working Group II to the Third Assessment Report of the Intergovernmental Panel on Climate Change. Cambridge: Cambridge University Press.

Schulze, R. 2005. Personal communication. 12 December.

Stainforth, D.A., Aina, T., Christensen, C., Collins, M., Faull, N., Frame, D.J., Kettleborough, J.A., Knight, S., Martin, A., Murphy, J.M., Piani, C., Sexton, D., Smith, L.A., Spicer, R.A., Thorpe, A.J. and Allen, M.R. 2005. Uncertainty in predictions of the climate response to rising levels of greenhouse gases. *Nature*, Vol. 433, 403–406.

Stott, P., Stone, D. and Allen, M. 2004. Human contribution to the European heat wave of 2003. *Nature*. Vol. 432: 610–614, December.

Tadross, M., Jack, C. and Hewitson, B. 2005. On RCM-based projections of change in southern African summer climate. *Geophysical Research Letters*. Vol. 32, L23713.

Thomas, C.D., Cameron, A., Green, R.E., Bakkenes, M., Beaumont, L.J., Collingham, Y.C., Erasmus, B., De Siqueira, M.F., Grainger, A., Hannah, L., Hughes, L., Huntley, B., Van Jaarsveld, A.S., Midgley. G.F., Miles, L., Ortega-Huerta, M.A., Townsend Peterson, A., Phillips, O.L. and Williams, S.E. 2004. Extinction risk from climate change. *Nature*. Vol. 427, 145–148.

United States Environmental Protection Agency. 2004. Greenhouse effect . . . [Online]. Available: http://www.epa.gov/globalwarming/kids/greenhouse.html. [Accessed 11 April 2005].

# Chapter 3

## Notes

1 Sections of this story appeared in *Africa Geographic*, April 2005.

2 Bond & Archibald, 2003: 381–389

3 Scholes, Midgley & Wand: 4

4 Bond: 2004

5 Scholes, Midgley & Wand: 2

6 Scholes, Midgley & Wand: 2

7 Scholes, Midgley & Wand: 3

8 Scholes, Midgley & Wand: 4

9 Scholes: 2005

10 'Elephants and Ecosystems', 2003: 1

11 'Elephants and Ecosystems', 2003: 1

12 'Interaction of Humans', 1998: 7

13 'Interaction of Humans', 1998: 6

14 'Elephants and Ecosystems', 2003: 3

15 *'Adansonia digitata'*, 2004: 1

16 'Ecosystem function modeling', 2005

17 'Ecosystem function modeling', 2005

## References

Bond, W. 2004. Interview with author on 27 February. Cape Town: University of Cape Town. [Transcript in possession of author].

Bond, W.J. and Archibald, S. 2003. Confronting complexity: fire policy choices in South African savanna parks. *International Journal of Wildland Fire*. 12(4): 381–389.

Bond, W.J. and Midgley, G.F. 2000. A proposed $CO_2$-controlled mechanism of woody plant invasion in grasslands and savannas. *Global Change Biology*. 6: 865–870.

Hankey, A. 2004. *Adansonia digitata A L*. [Online.] Available: www.plantzafrica.com/plantab/adansondigit.html. [Accessed February 2005.]

National Botanical Institute. 2001. The heat is on: impacts of climate change on plant diversity in South Africa. [Brochure].

Owen-Smith, N. 1998. Presentation: The interaction of humans, megaherbivores, and habitats in the late Pleistocene extinction event. [Online]. The American Museum of Natural History. Available: http://www.amnh.org/science/biodiversity/extinction/Day1/bytes/SmithPres.html. [Accessed March 2005].

Owen-Smith, N. 2003. Elephants and ecosystems. For proceedings of the Norwegian Bonic Programme Workshop, Kasane, Botswana, March. [Digital copy of notes in possession of author].

Scholes, B. 2005. Ecosystem function modeling. AIACC Biodiversity Course Notes in PowerPoint format. [Digital copy in possession of author].

Scholes, B. 2005. Telephonic interview with author in March. [Handwritten notes with author].

Scholes, R J., Midgley, G.F. and Wand, S.J.E. 2000. The vulnerability and adaptation of rangelands. In: Kiker, G. Climate change impacts in Southern Africa. Report to the National Climate Change Committee, Department of Environmental Affairs and Tourism, Pretoria, South Africa.

# Chapter 4

## Notes

1 'Vegetation of southern Africa', no date; and Cowling & Richardson, 1995: 31

2 Cowling & Richardson, 1995: 7

3 'Cape flora on World Heritage list', 2004

4 Pauw & Johnson, 1999: 17

5 'Fynbos background', 2005

6 Cowling & Richardson, 1995: 7

7 'Fynbos biome', no date

8 Cowling & Richardson, 1995: 59

9 Cowling & Richardson, 1995: 58

10 Cowling & Richardson, 1995: 43

11 Cowling & Richardson, 1995: 48

12 Cowling & Richardson, 1995: 49

13 Cowling & Richardson, 1995: 73

14 Brown, 1993

15 Cowling & Richardson, 1995: 60

16 Cowling & Richardson, 1995: 7

17 Midgley et al., 2005: 31

18 Midgley et al., 2005: 15

19 Midgley et al., 2005: 28

20 Simmons et al., 2004: 299

21  Heath & Brinkman, 1995

22  Heath, 2005

23  Claassens, A. 2004; and Heath & Brinkman, 1995; and Heath, 1998

24  Midgley et al., 2002: 446

25  Simmons et al., 2004: 297

26  Midgley et al., 2005: 28

27  Pauw & Johnson, 1999: 38

28  Pauw & Johnson, 1999: 52


## References

Brown, N.A.C. 1993. Promotion of germination of fynbos seeds by plant-derived smoke. *New Phytologist*. Vol. 123, No. 3, March: 575–583.

Claassens, A. 2004. Interview with author in February, Cape Town. [Audio and printed copies in possession of author].

Cowling, R. and Richardson, D. 1995. *Fynbos: South Africa's Unique Floral Kingdom*. Cape Town: University of Cape Town.

Flower Valley Conservation Trust. 2005. Fynbos background. [Online]. Available: http://www.flowervalley.org.za/f_1024.htm. [Accessed January 2005].

Heath, A. 1998. Further aspects on the life history of the myrmecophilous species *Chrysoritis dicksoni* (Gabriel), (Lepidoptera: Lycaenidae). *Metamorphosis*. Vol. 9, No. 4, December.

Heath, A. 2005. Personal email correspondence with author. [Digital copy in possession of author].

Heath, A. and Brinkman, A.K. 1995. Aspects of the life history, distribution and population fluctuations of *Oxychaeta dicksoni* (Gabriel) (Lepidoptera: Lycaenidae). *Metamorphosis*. Vol. 6, No. 3, September.

Midgley, G.F., Chapman, R.A., Hewitson, B., Johnston, P., de Wit, M., Ziervogel, G., Mukheibir, P., van Niekerk, L., Tadross, M., van Wilgen, B.W., Kgope, B., Morant, P.D., Theron, A., Scholes, R.J. and Forsyth, G.G. 2005. *A Status Quo, Vulnerability and Adaptation Assessment of the Physical and Socio-economic Effects of Climate Change in the Western Cape*. Report to the Western Cape Government, Cape Town, South Africa. CSIR Report No. ENV-S-C 2005-073, Stellenbosch.

Midgley, G.F., Hannah, L., Millar, D., Rutherford, M.C. and Powrie, L.W. 2002. Assessing the vulnerability of species richness to anthropogenic climate change in a biodiversity hotspot. *Global Ecology & Biogeography*. 11: 445 – 451.

National Botanical Institute. 2001. The heat is on: Impacts of climate change on plant diversity in South Africa. [Brochure].

National Botanical Institute. No date. Fynbos biome. [Online]. Available: http://www.plantzafrica.com/vegetation/fynbos.htm. [Accessed June 2004].

National Botanical Institute. No date. Fynbos: South Africa's unique floral kingdom. [Online]. Available: http://www.plantzafrica.com/vegetation/vegmain.htm. [Accessed March 2004].

National Botanical Institute. No date. Vegetation of southern Africa. [Online]. Available: http://www.plantzafrica.com/vegetation/vegmain.htm. [Accessed March, June 2004].

Pauw, A. and Johnson, S. 1999. *Table Mountain: A Natural History*. Cape Town: Fernwood Press.

Simmons, R.E., Barnard, P., Dean, W.R.J., Midgley, G.F., Thuiller, W. and Hughes, G. 2004. Climate change and birds: perspectives and prospects from southern Africa. *Ostrich*. 75(4): 295–308.

SouthAfrica.info. 2004. Cape flora on World Heritage list. [Online]. Available: http://www.southafrica.info/ess_info/sa_glance/fauna_flora/capefloralregion.htm. [Accessed January 2005].

## Chapter 5

### Notes

1   Minter, 2005

2   Minter et al., 2004: 75

3   Minter et al., 2004: 204

4   Burnier, 2004: 86

5   Burnier, 2004: 430

6   Burnier, 2004: 430

7   Burnier, 2004: 430

8   Burnier, 2004: 364

9   Minter et al., 2004: 95

10  Minter et al., 2004: 95

11  Minter et al., 2004: 108

12  Minter et al., 2004: 87

13  Minter et al., 2004: 33

14  Minter et al., 2004: 33–34

15  Minter et al., 2004: 33–34

16  Minter et al., 2004: 11

17  Minter et al., 2004: 89

18  Minter et al., 2004: 26

19  Minter et al., 2004: 28

### References

Burnier, D. (Editor-in-chief). 2004. *Animal: The Definitive Visual Guide*. London: Dorling Kindersley.

De Villiers, A. 2004. Interview with author on 8 June. [Handwritten notes in possession of author].

Houghton, J.T., Ding, Y., Griggs, D.J., Noguer, M., Van der Linden, P.R., Dai, X., Maskell, K. and

Johnson, C.A. (Eds). 2001. *Climate Change 2001: The Scientific Basis.* Contribution of Working Group 1 to the Third Assessment Report of the Intergovernmental Panel on Climate Change. Cambridge: Cambridge University Press.

Minter, L.R. 2005. Telephonic interview with author on 23 March. [Handwritten notes in possession of author].

Minter, L.R., Burger, M., Harrison, J.A., Braack, H.H., Bishop, P.J. and Kloepfer, D. (Eds). 2004. *Atlas and Red Data Book of the Frogs of South Africa, Lesotho and Swaziland.* SI/MAB Series #9. Washington, DC: Smithsonian Institute.

National Botanical Institute. 2001. The heat is on: Impacts of climate change on plant diversity in South Africa. [Brochure].

Turner, A. 2004. Interview with author on 8 June. [Handwritten notes in possession of author].

# Chapter 6

## Notes

1   Silcock, 1957

2   Underhill, 2005: 'while the Curlew yearlings don't fly to the tundra for their first winter, the little stilts do'

3   'The mystery of migration', 2002

4   'The mystery of migration', 2002

5   Underhill, 2005

6   Harebottle & Underhill, 2003

7   Burnier, 2004: 303

8   'Skyhawk specifications', 2005

9   William & Williams, 1978: 166–176

10  Cooper, 2003

11  Underhill, 2005

12  'Bird behaviour', 2003; and 'How birds fly', 2001

13  Simmons et al., 2004: 297

14  Schekkerman et al., 2004: 20; and Romanovsky, 2005: 20

15  Schekkerman et al., 2004: 10

16  Romanovsky, 2005: 19

17  Schekkerman et al., 2004: 77

18  Underhill, 2005

19  Harebottle, 2005

20  Rehfisch & Crick, 2003: 90

21  Rehfisch & Crick, 2003: 90

22  Rehfisch & Crick, 2003: 88

23  Simmons et al., 2004: 299

24  Rehfisch & Crick, 2003: 88

25  Rehfisch & Crick, 2003: 89

26  Rehfisch & Crick, 2003: 89

27  Harebottle et al., 2003

28  Rehfisch & Crick, 2003: 86–100

## References


BBC.co.uk. 2002. Science & Nature: Animals: The mystery of migration. [Online]. Available: http://www.bbc.co.uk/nature/animals/birds/weeklyfeature/migration/. [Accessed March 2005].

Burnier, David (Editor-in-chief). 2004. *Animal: The Definitive Visual Guide*. London: Dorling Kindersley.

Cessna.com. 2005. Skyhawk specifications & description. [Online]. Available: http://skyhawk.cessna.com/spec_perf.chtml. [Accessed February 2005].

Cooper, J. 2003. Interview with the author in March. Marion Island. [Digital audio copy in possession of author].

Harebottle, D.M., Navarro, R.A., Underhill, L.G. and Waltner, M. 2003. Trends in numbers of migrant waders (*Charadrii*) at Langebaan Lagoon, South Africa, 1975–2003. [Poster]. [Digital copy in possession of the author].

Harebottle, D.M. and Underhill, L.G. 2003. The Arctic connection: monitoring coastal waders in South Africa, a case study. [Poster]. [Digital copy in possession of the author].

http://nationalzoo.si.edu/ConservationAndScience/MigratoryBirds/Fact_Sheets/default.cfm?fxsht=4. [Accessed February 2005].

Hughes, G. 2004. Climate change and birds: perspectives and prospects from southern Africa. *Ostrich*, 75(4): 295–308.

Luhr, J.F. (Ed). 2004. *Earth: The Definitive Visual Guide*. London: Dorling Kindersley.

Rehfisch, M.M. and Crick, H.Q.P. 2003. Predicting the impact of climatic change on Arctic-breeding waders. *Wader Study Group Bulletin*, 100: 86–100. April.

Silcock, A. 1957. *Verse and Worse: Private Collection by Arnold Silcock*. London: Faber & Faber.

Simmons, R.E., Barnard, P., Dean, W.R.J., Midgley, G.F., Thuiller, W., Schekkerman, H., Tulp, I., Calf, K.M. and De Leeuw, J.J. 2004. *Studies on Breeding Shorebirds at Medusa Bay, Taimyr, in Summer 2002*. Netherlands: Wageningen.

Sturm, M., Schimel, J., Michaelson, G., Welker, J.M., Oberbauer, S.F., Liston, B.E., Fahnestock, J. and Romanovsky, V.E. 2005. Winter biological processes could help convert Arctic tundra to shrubland. *Bioscience*. 55(1): 7–9, January.

Transport Canada Civil Aviation. 2003. Bird behaviour that may create aviation hazards. [Online]. Available: http://www.tc.gc.ca/CivilAviation/Aerodrome/WildlifeControl/tp13549/Chapter3/Chapter3h.htm. [Accessed February 2005].

Underhill, L. 2005. Interview with the author on 24 February, Cape Town. [Digital and handwritten copy in possession of author].

William, T.C. and Williams, J.M. 1978. An oceanic mass migration of land birds. *Scientific American*. 239(4): 166–176. In Deinlein, M. No date. Smithsonian National Zoological Park: Migratory bird centre: Have wings, will travel: Avian adaptations to migration. [Online]. Available:http://nationalzoo.si.edu/ConservationandScience/Migratory/Birds/Fact_Sheets/default.cfm?fxsht=4

# Chapter 7

## Notes

1  Wilson, 2003: 93

2  Wilson, 2003: 94

3  Wilson, 2003: 93

4  Wilson, 2003: 92

5  Wilson, 2003: 94

6  Quammen, 1996: 54

7  Flynn & Wyss, 2002: 57–58

8  Flynn & Wyss, 2002: 62

9  Flynn & Wyss, 2002: 62

10 Wilson, 2003: 93–94

11 'Island biogeography', 1988

12 Cowling & Richardson, 1995: 32–33

## References

Botanic Society. 2002. Lowlands flora. [Online]. Available: http://www.botanicalsociety.org.za/CCU/downloads/articles/Veld&Flora_LowlandsFLORA_Dec02.PDF. [Accessed 2004].

Cowling, R. and Richardson, D. 1995. *Fynbos: South Africa's Unique Floral Kingdom*. Cape Town: University of Cape Town.

Ehrlich, P.R., Dobkin, D.S. and Wheye, D. 1988. Island biogeography. [Online]. Available: http://www.stanfordalumni.org/birdsite/text/essays/Island_Biogeography.html. [Accessed December 2003].

Flynn, J.J. and Wyss, A.R. 2002. Madagascar's Mesozoic secrets. *Scientific American*. February.

Hitchcock, A. 2003. *Erica jasminiflora*. [Online]. Available: http://www.plantzafrica.com/plantefg/ericajasmin.htm. [Accessed 24 April 2005].

Pauw, A. and Johnson, S. 1999. *Table Mountain: A Natural History*. Cape Town: Fernwood Press.

Quammen, D. 1996. *The Song of the Dodo*. London: Pimlico, Random House.

Simmons, R.E., Barnard, P., Dean, W.R.J., Midgley, G.F., Thuiller, W. and Hughes, G. 2004. Climate change and birds: perspectives and prospects from southern Africa. *Ostrich*, 75(4): 295– 308.

Wilson, E.O. 2003. *The Future of Life*. London: Abacus.

# Chapter 8

## Notes

1  Burnier, 2004: 366

2  'Western Cape's tortoises', no date

3  Leslie & Spotila, 2001: 347–355

## References


Burnier, D. (Editor-in-chief). 2004. *Animal: The Definitive Visual Guide*. London: Dorling Kindersley.

Cape Nature Conservation. No date. The Western Cape's tortoises. [Brochure]. [Photocopy in possession of the author].

Henen, B.. 2003. Interview with the author in October.

Leslie, A.J. and Spotila, J.R. 2001. Alien plant threatens Nile crocodile (*Crocodylus niloticus*) breeding in Lake St Lucia, South Africa. *Biological Conservation*. 98(2001), 347–355.

Leuteritz, T. 2003. Interview with the author in September, Prince Albert. [Digital audio and typed copies in possession of the author].


# Chapter 9

## Notes

1  Thanks to Charles Dodgson (otherwise known as Lewis Carroll) for the masterful nonsense of *The Jobberwocky*. Apologies for the corruption thereof.

2  Luhr, 2004: 383–391

3  Luhr, 2004: 127

4  By way of comparison, the West Coast's mean summer sea surface temperature is 14°C with winter at 15°C; on the East Coast it is 27°C in summer and 22°C in winter.

5  Cockroft, 2001: 1088

6  Cockroft, 2001: 1086

7  Clark, 2005: 84–95

8  Clark et al., 2000

9  Tarr et al., 1996: 319–323

## References


Clark, B.M. 2006. Climate change: A looming challenge for fisheries management in southern Africa. *Marine Policy*. 30(1): 84–95.

Clark, B.M., Steffani, N.C., Young, S., Richardson, A.J. and Lombard, A.T. 2000. The effects of climate change on marine biodiversity in South Africa. In: Kiker, G. Climate change impacts in Southern Africa. Report to the National Climate Change Committee, Department of Environmental Affairs and Tourism, Pretoria, South Africa.

Cockcroft, A.C. 2001. *Jasus lalandii* 'walkouts' or mass strandings in South Africa during the 1990s: an overview. *Marine and Freshwater Research*. 52(8): 1085–1093.

Hutchings, L. 2004. Personal communications. [Digital copy in possession of author].

Peters, R.L. and Lovejoy, T.E. 1992. *Global Warming and Biological Diversity*. Yale University Press: 91–99.

Schumann, E.H., Cohen, A.L. and Jury, M.R. 1995. Coastal sea surface temperature variability along the south coast of South Africa and the relationship to regional and global climate. *Journal of Marine Research*. 53: 231–248.

Tarr, R.J.P., Williams, P.V.G. and MacKenzie, A.J. 1996. Abalone, sea urchins and rock lobsters: a possible ecological shift may affect traditional fisheries. *South African Journal of Marine Science*. 17: 319–323.

Turpie, J.K., Heydenrych, B.J. and Lamberth, S.J. 2003. Economic value of terrestrial and marine biodiversity in the Cape Floristic Region: implications for defining effective and socially optimal conservation strategies. *Biological Conservation*. 112: 233–251.

# Chapter 10

## Notes

1 'The top ten coral reef hotspots', 2002

2 Wilkinson & Buddemeier, 1994: 12

3 'The top ten coral reef hotspots identified', 2002

4 Wilkinson & Buddemeier, 1994: ix

5 Hoegh-Guldberg, 1999: 841

6 Hoegh-Guldberg, 1999: 841

7 Hoegh-Guldberg, 1999: 839

8 Hoegh-Guldberg, 1999: 848

9 Hoegh-Guldberg, 1999: 840

10 Hoegh-Guldberg, 1999: 853

11 Hoegh-Guldberg, 1999: 844

12 Hoegh-Guldberg, 1999: 844

13 Obura, 2005

14 Schleyer & Cilliers, **2003:** 387

15 Schleyer & Cilliers, **2003:** 392

16 Schleyer & Cilliers, **2003:** 397

17 Schleyer & Cilliers, **2003:** 395

18 Kleypas et al., 1999: 157

19 Hoegh-Guldberg, 1999: 839

20 Hoegh-Guldberg, 1999: 858

21 Hoegh-Guldberg, 1999: 858

22  Hoegh-Guldberg, 1999: 858

23  Hoegh-Guldberg, 1999: 858

24  Pomerance, 1999

25  Munro, 1996

26  Bryant et al., 1998

## References


Brown, B.E. 1997. Coral bleaching: causes and consequences. *Coral Reefs*. 16: 129–138, Spring.

Bryant, D., Burke, L., McManus, J. and Spalding, M. 1998. Reefs at risk: a map-based indicator of threats to the world's coral reefs. Washington, D.C.: World Resources Institute.

Conservation International. 2002. Press releases: The top ten coral reef hotspots. [Online]. Available: http://www.conservation.org/xp/news/press_releases/2002/021402a.xml. [Accessed May 2004].

Hoegh-Guldberg, O. 1999. Climate change, coral bleaching and the future of the world's coral reefs. *Marine and Freshwater Research*. 50(8): 839–866.

Kleypas, J.A., McManus, J.W. and Menez, L.A.B. 1999. Environmental limits to coral reef developments: Where do we draw the line? *American Zoologist*. 39: 146–159, February.

Munro, J. L. 1996. The scope of tropical reef fisheries and their management. In Polunin, N.V.C. and Roberts, C.M. *Reef Fisheries*. London: Chapman & Hall: 1–14.

Obura, D. 2005. Climate change and African coral reefs. Paper presented at the National Climate Change Conference, Gauteng.

Pearce, F. 2002. NewScientists.com: The top ten coral reef hotspots identified. [Online]. Available: http://www.newscientist.com/article.ns?id=dn1927. [Accessed May 2004].

Pomerance, R. 1999. Coral bleaching, coral mortality, and global climate change. Report presented by the Deputy Assistant Secretary of State for the Environment and Development to the US Coral Reef Task Force, 5 March, Maui, Hawaii.

Schleyer, M. and Celliers, L. 2003. Biodiversity on marginal coral reefs of South Africa: What does the future hold? *Zool. Verh. Leiden*: 345, 387–400.

Thande, George. 2004. Reuters: Indian Ocean could lose coral islands in 50 years. [Online]. Available: http://www.nzherald.co.nz/section/story.cfm?c_id=2&objectid=3565632. [Accessed November 2004].

Wilkinson, C.R. and Buddemeier, R.W. 1994. Global climate change and coral reefs: implications for people and reefs. Report of the UNEP-IOC-ASPEI-IUCN Global Task Team on the Implications of Climate Change on Coral Reefs. Gland, Switzerland: IUCN (The World Conservation Union).


# Chapter 11

## Notes

1  'The heat is on', 2001: 5

2   Van Jaarsveld, 2004

3   'Aloe dichotoma', 2002

4   Musil et al., 2004: 1

5   Musil et al., 2004: 6

6   'The heat is on', 2001: 5

## References


Foden, W. 2004. Interview with the author on 28 January, Cape Town. [Digital audio and printed typed copies in possession of author].

Houghton, J.T., Ding, Y., Griggs, D.J., Noguer, M., Van der Linden, P.R., Dai, X., Maskell, K. and Johnson, C.A (Eds). 2001. *Climate Change 2001: The Scientific Basis.* Contribution of Working Group I to the Third Assessment Report of the Intergovernmental Panel on Climate Change. Cambridge: Cambridge University Press.

Milton, S.J., Yeaton, R.I., Dean. W.R.J. and Vlok, J.H.I. 1997. Succulent karoo. In Cowling, R.M., Richardson, D.M. and Pierce, S.M. (Eds). *Vegetation of Southern Africa.* Cambridge: Cambridge University Press, 131–161.

Musil, C.F., Schmiedel, U. and Midgley, G.F. 2004. Lethal effects of experimental warming approximating a future climate scenario on southern African quartz-field succulents: a pilot study. *New Phytologist.*

National Botanical Institute. 2001. The heat is on: Impacts of climate change on plant diversity in South Africa. [Brochure].

National Botanical Institute. 2002. *Aloe dichotoma* Masson. [Online]. Available: http://www.plantzafrica.com/plantab/aloedichotoma.htm. [Accessed April 2004].

National Botanical Institute. No date. Floral kingdoms. [Online]. Available: http://www.plantzafrica.com/vegetation/floralkingdoms.htm. [Accessed April 2004].

*Reader's Digest Illustrated History of South Africa, The Real Story.* (3rd Rev. Ed.). 1994. Cape Town: Reader's Digest.

Simmons, R.E., Barnard, P., Dean, W.R.J., Midgley, G.F., Thuiller, W. and Hughes, G. 2004. Climate change and birds: perspectives and prospects from southern Africa. *Ostrich*, 75(4): 295–308.

Van Jaarsveld, E. 2004. Interview with the author on 15 January, Cape Town. [Digital audio and typed copies held in possession of author].

Wildworld Full Report. 2001. Montane fynbos and renosterveld. [Online]. Available:http://www.worldwildlife.org/wildworld/profiles/terrestrial/at/at1203_full.html. [Accessed April 2004].


# Chapter 12

## Notes

1   Marsh, 1948

2   Marsh, 1948

3   Marsh, 1948

4   Luhr, 2004: 384

5   Bester, 2005

6   Hanel & Chown, no date

7   Crick, 2004: 52

## References


Bester, Marthan. 2005. Interview with the author on 18 March, Stellenbosch. [Notes in possession of author].

Crick, H. 2004. The impact of climate change on birds. *Ibis.* 146 (Suppl.1): 48–56.

Hanel, C. and Chown, S. No date. *An Introductory Guide to the Marion and Prince Edward Islands Special Nature Reserves 50 years after Annexation*. South Africa: Department of Environmental Affairs and Tourism, Directorate Antarctica and Islands.

Luhr, J.F. (Ed). 2004. *Earth: The Definitive Visual Guide*. London: Dorling Kindersley.

Marsh, J.H. 1948. *No Pathway Here*. Cape Town: The Standard Press; London: Hodder & Stoughton.

McGeoch, M. 2003. Interview with the author in April. [Transcript in possession of author].

Meldrum, A. and Kelso, P. 2005. Cyber Diver News Network: Deep death, extreme diver Dave Shaw pays the ultimate price. [Online]. Available:http://www.cdnn.info/news/safety/s050114.html. [Accessed 6 April 2005].


# Chapter 13

## Notes

1   Ponting, 2001: 495

2   Ponting, 2001: 411

3   Ponting, 2001

4   The original text refers to 5 feet 7 inches.

5   Bryson & Murray, 1977; and Herman, 1954. Here the original text refers to 'less than five feet tall'.

6   Bryson & Murray, 1977; and Herman, 1954

7   Ponting, 2001: 24–25

8   Ponting, 2001: 34

9   Ponting, 2001: 52

10 Ponting, 2001: 53

11 Ponting, 2001: 52

12 Ponting, 2001: 447

13 Couper-Johnston, 2000: 110

14 Couper-Johnston, 2000: 111

15 Couper-Johnston, 2000: 111–112

16 'City statistics', 2003

17 Couper-Johnston, 2000: 14

18 Couper-Johnston, 2000: 10

19 Ponting, 2001: 791

20 'Portrait of a famine', 2000

21 Ponting, 2001; and 'The Soviet famines', 2005

22 Ponting, 2001: 791; and 'Hungry ghosts', 1996

23 Lomborg, 1998: 63

24 'Biodiversity food for security', 2004

25 'Enough food in the future', 2004

26 'Biodiversity food for security', 2004

27 Ponting, 2001: 791

28 'Red Cross recognizes World Food Day 2004': 2004

29 According to Lomborg, improvements in irrigation and fertilizer use, which have increased agricultural output in Asia, have not taken off to the same degree in Africa.

30 Lomborg, 1998: 66

31 'The dilemma of famine', 2003

32 'The state of food insecurity', 2003

33 Schneider et al., 2001: 84

34 McMichael & Githeko, 2001: 473

35 'South Africa: Initial National Communication', 2000: 35

37 'South Africa: Initial National Communication', 2000: 35

37 'South Africa: Initial National Communication', 2000: 38

38 Du Toit et al., 2000: 2

39 Sprattmoron, 2003

40 'Living beyond our means', 2005: 24

## References


African Centre for Disaster Studies. 2004. Red Cross recognizes World Food Day 2004. [Online]. Available: http://acds.co.za/index.php?myname=News&storyid=24&searcher=. [Accessed 7 June 2005].

Associated Press. 2003. *Guardian Unlimited:* Global warming could trigger mass extinction. [Online]. Available: www.guardian.co.uk/climatechange/story/0,12374,980561,00.html. [Accessed April 2005].

Becker, J. 1996. *Hungry Ghosts: Mao's Secret Famine.* New York: Henry Holt & Company. [Online]. Available: http://www.overpopulation.com/faq/health/hunger/famine/chinese_famine.html.

Biodiversity: our food depends on it. 2004. Biodiversity food for security. [Online]. Available: http:/ /www.fao.org/wfd/index_en.asp. [Accessed March 2005].

Bryson, B. 2003. *A Short History of Nearly Everything*. London: Doubleday.

Bryson, R.A. and Murray, T.J. 1977. Madison, Wesconsin: University of Wesconsin Press.

Couper-Johnston, R. 2000. *El Niño, the Weather Phenomenon that Changed the World*. London: Hodder & Stoughton.

Desanker, P. and Magadza, C. 2001. Africa. In McCarthy, J.J., Canziani, O.F., Leary, N.A., Dokken, D.J. and White, K.S. (Eds). 2001. *Climate Change 2001: Impacts, Adaptation, and Vulnerability*. Contribution of Working Group II to the Third Assessment Report of the Intergovernmental Panel on Climate Change. Cambridge University Press: Cambridge.

Du Toit, A.S., Prinsloo, M.A., Durand W. and Kiker, G. 2000. Vulnerability of maize production to climate change and adaptation assessment in South Africa. In: Kiker, G. Climate change impacts in Southern Africa. Report to the National Climate Change Committee, Department of Environmental Affairs and Tourism, Pretoria, South Africa.

Food and Agriculture Organization of the United Nations. 2003. The state of food insecurity in the world: monitoring progress towards the World Food Summit and Millennium Development Goals. [Online]. Available: http://www.fao.org. [Accessed February 2005].

Herman, P. 1954. *Conquest of Man*. New York: Harper.

Houghton, J.T., Ding, Y., Griggs, D.J., Noguer, M., Van der Linden, P.R., Dai, X., Maskell, K. and Johnson, C.A (Eds). 2001. *Climate Change 2001: The Scientific Basis*. Contribution of Working Group I to the Third Assessment Report of the Intergovernmental Panel on Climate Change. Cambridge: Cambridge University Press.

Humanitarian Aid. 2005. The Soviet famines of 1921 and 1932–3. [Online]. Available: http:// www.overpopulation.com/faq/health/hunger/famine/soviet_famine.html. [Accessed March 2005].

Jo'burg: City of Johannesburg official website. 2003. City statistics: Joburg according to the census, 1996–2001. [Online]. Available: http://www.joburg.org.za/corporate_planning/ corpplan_citystats.stm. [Accessed March 2005].

Lamb, H.H. 1966. *Climate, History and the Modern World*. London: Methuen.

Lomborg, B. 1998. *The Skeptical Environmentalist: Measuring the Real State of the World*. Cambridge: Cambridge University Press.

McMichael, A. and Githeko, A. 2001. Human health. In McCarthy, J.J., Canziani, O.F., Leary, N.A., Dokken, D.J. and White, K.S. (Eds). 2001. *Climate Change 2001: Impacts, Adaptation, and Vulnerability*. Contribution of Working Group II to the Third Assessment Report of the Intergovernmental Panel on Climate Change. Cambridge University Press: Cambridge.

Millennium Ecosystem Assessment. 2005. Living beyond our means: natural assets and human well-being. [Online]. Available: www.milleniumassessment.org. [Accessed 4 April 2005]. [Digital copy in possession of author].

Millennium Ecosystem Assessment. 2005. Millennium Ecosystem Assessment Synthesis Report. [Online]. Available: www.milleniumassessment.org. [Accessed 4 April 2005]. [Digital copy in possession of author].

Milner, K. 2000. Flashback 1984: Portrait of a famine. [Online]. Available: http://news.bbc.co.uk/1/hi/world/africa/703958.stm. [Accessed March 2005].

Monbiot, G. 2003. *Guardian Unlimited:* Shadow of extinction. July 1. [Online]. Available: www.guardian.co.uk/comment/story/0,3604,988380,00.html. [Accessed July 2003].

National Botanical Institute. 2001. The heat is on: Impacts of climate change on plant diversity in South Africa. [Brochure].

Ponting, C. 2001. *World History: A New Perspective.* London: Pimlico.

Riddley, K.J. 2003. The dilemma of famine and drought in Africa. [Online]. Available: http://www.africasummit.org/news/krista.pdf. [Accessed February 2005].

Schneider, S. and Sarukhan, J. 2001. Overview of impacts, adaptation, and vulnerability to climate change. In McCarthy, J.J., Canziani, O.F., Leary, N.A., Dokken, D.J. and White, K.S. (Eds). 2001. *Climate Change 2001: Impacts, Adaptation, and Vulnerability.* Contribution of Working Group II to the Third Assessment Report of the Intergovernmental Panel on Climate Change. Cambridge: Cambridge University Press.

Sprattmoran, S. 2003. *Foreword* magazine, March. Book review quoting a BBC interview with Jung in 1959. Available: http://www.forewordmagazine.com/reviews.aspx?reviewID=1008

Union of Concerned Scientists. 2004. FAO Report: Enough food in the future without genetically engineered crops. [Online]. Available: http://www.ucsusa.org/food_and_environment/biotechnology/page.cfm?pageID=331. [Accessed March 2005].

United Nations Framework Convention on Climate Change. 2000. South Africa: Initial national communication under the United Nations Framework Convention on Climate Change. [Online]. Available: http://unfccc.int/2860.php. [Accessed November 2004]. [Digital copy in possession of author].

# Chapter 14

## Notes

1 'Living beyond our means', 2005: 23

2 'Millennium Ecosystem Assessment Synthesis Report', 2005: 62

3 Wilson, 2002: 106

4 Freeman, 2002: 2

5 'Living beyond our means', 2005: 4

6 Radford, 2005: 28

7 Wilson, 2002: 140

8 Wilson, 2002: 140

9 Wilson, 2002: 133

10 Wilson, 2002: 132

11 'Living beyond our means', 2005: 17

12 Turpie et al., 2003: 233–251

13 Bryson, 2003: 303

14 Benton, 2003

15 Monbiot, 2003

16 'Heat will soar', 2003

17 Allen & Lord, 2004: 535–655

18 Black, 2005; and 'Climate change', 2005

19 United Nations Framework Convention on Climate Change, no date

20 Clarke, 2002: 52

21 'Global warming', 2005

22 'CAIT', 2005

23 Ponting, 2000: 800

24 Ponting, 2000: 801

25 Roberts, 2004: 38

26 Gore, 2005

27 Phillips, 2004

28 Cave, 2005

29 Roberts, 2004: 94

30 Mann et al., 1998: 779–787

31 Townsend & Harris, 2004

32 Townsend & Harris, 2004

33 Monbiot, 2004

34 Brown et al., May 2005

35 Liu & Diamond, 2005: 1179–1186

36 Aldhous, 2005: 1152–1154

37 Aldhous, 2005: 1152–1154

38 Richter et al., 2005: 129–132

39 Liu & Diamond, 2005: 1179–1186

40 Liu & Diamond, 2005: 1179–1186

41 These figures, from the World Energy Institute, are for the year 2000 and are regarded as the most up to date.

42 Sparks & Mwakasonda, 2004: 101

43 'CAIT', 2005

44 Kenny, 2004: 54

45 Winkler, 2004: 25

46 Nkomo, 2004: 84

47 Sparks & Mwakasonda, 2004: 95

48 'US in race to unlock new energy source', 2005

49 Wilson, 2003: 23

50 'Living beyond our means', 2005: 24

51 Bryson, 2003: 303; and 'Global warming could trigger mass extinction', 2003

## References


Aldhous, P. 2005. China's burning ambition. *Nature*. 435: 1152–1154

Allen, M.R. and Lord, R. 2004. The blame game: Who will pay for the damaging consequences of climate change? *Nature*. 432(7017): 535–655, December 2.

Associated Press. 2003. *Guardian Unlimited:* Global warming could trigger mass extinction. June 19. [Online]. Available: www.guardian.co.uk/climatechange/story/0,12374,980561,00.html. [Accessed April 2005].

BBC News. 2005. On this Day, 11 November 1987: Van Gogh fetches record price. [Online]. Available: http://news.bbc.co.uk/onthisday/hi/dates/stories/november/11/newsid_2539000/ 2539613.stm. [Accessed 5 April 2005].

Benton, M. 2003. *When Life Nearly Died: The Greatest Mass Extinction of All Time*. London: Thames & Hudson.

Black, R. 2005. BBC News: Alarm at new climate warning. January 26. [Online]. Available: http:// news.bbc.co.uk/1/hi/sci/tech/4210629.stm. [Accessed February 2005].

Brown, P., Wintour, P. and White, M. 2005. Leaked G8 draft angers green groups. May. [Online]. Available: http://politics.guardian.co.uk/green/story/0,9061,1494434,00.html. [Accessed 28 May 2005].

Bryson, B. 2003. *A Short History of Nearly Everything*. London: Doubleday.

Cave, D. 2005. The United States of Oil. Salon.com. [Newspaper, selected stories online]. Retrieved 17 November 2005 from World Wide Web: www.salon.com/tech/feature/2001/11/19/ bush_oil/index.html.

Clarke, J. 2002. *Coming Back to Earth: South Africa's Changing Environment*. Houghton: Jacana.

Cyranoski, D. 2005. Satellite view alerts China to soaring pollution. *Nature*. 437: 12.

Davidson, O. and Winkler, H. 2004. Energy policy. In Energy for Sustainable Development – South African: Profile Phase 1 Final Report. Energy Research Centre, University of Cape Town. Unpublished.

Food and Agriculture Organization of the United Nations. 2004. FAO newsroom: World Food Day 2004 highlights the importance of biodiversity to global food security. October 15. [Online]. Available: http://www.fao.org/newsroom/en/news/2004/51140/index.html. [Accessed January 2005].

Freeman, A.M. 2002. Valuing nature: a set of papers resulting from the Shipman workshop, Bowdoin College. [Online]. Available: http:// academic.bowdoin.edu/environmental_studies/ dissemination/valnat.pdf. [Accessed 18 May 2005].

Gore, A. 2005. It is no longer possible to ignore the strangeness of our public discourse. Boston.com. [Newspaper, selected stories online]. Retrieved 22 October 2005 from World Wide Web: www.boston.com/news/nation/articles/2005/10/06 text_of_gore_speech_at_media_conference/.

Greenpeace. No date. George W Bush. [Online]. Available: http://www.greenpeace.org.uk/climate/climatecriminals/bush/index.cfm. [Accessed 21 April 2005].

*Guardian Unlimited*. 2005. US in race to unlock new energy source. [Online]. Available: http://www.mg.co.za/articledirect.aspx?articleid=200777&area=%2finsight%2finsight_international%2f). [Accessed 21 April 2005].

Kenny, A. 2004. Energy supply in South Africa. In Energy for Sustainable Development – South African: Profile Phase 1 Final Report. Energy Research Centre, University of Cape Town. Unpublished.

Klopper, G. 2002. The fortune son. *The Big Issue*. Issue 65, Vol. 6: December.

Monbiot, G. 2004. War x 4. [Online]. Available: www.monbiot.com. [Accessed 21 April 2005].

Lean, G. 2005. *HealthLink*: Linking health and the environment. Global warming approaching point of no return, warns leading climate expert. January 23. [Online]. Available: http://www.citizensinaction.org/climatechange42.html; also: http://news.independent.co.uk/world/environment/story.jsp?story=603752. [Accessed 4 April 2005].

Liu, J. and Diamond, J. 2005. China's environment in a globalizing world. *Nature*. 435: 1179–1186.

Mann, M.E., Bradley, R.S. and Hughes, M.K. 1998. Global-scale temperature patterns and climate forcing over the past six centuries. *Nature*. Vol. 392, 779–787.

Millennium Ecosystem Assessment. 2005. Living beyond our means: natural assets and human well-being. [Online]. Available: www.milleniumassessment.org. [Accessed 4 April 2005]. [Digital copy in possession of author].

Millennium Ecosystem Assessment. 2005. Millennium Ecosystem Assessment Synthesis Report. [Online]. Available: www.milleniumassessment.org. [Accessed 4 April 2005]. [Digital copy in possession of author].

Monbiot, G. 2003. *Guardian Unlimited:* Shadow of extinction. July 1. [Online]. Available: www.guardian.co.uk/comment/story/0,3604,988380,00.html. [Accessed July 2003].

NewScientist.com. 2005. Editorial: Climate change threat may be underestimated. February 12. [Online from issue 2486 of *NewScientist*]. Available: http://www.newscientist.com/channel/earth/climate-change/mg18524863.400. [Accessed 5 April 2005].

Nkomo, J.C. 2004. Energy and economic development. In Energy for Sustainable Development – South African: Profile Phase 1 Final Report. Energy Research Centre, University of Cape Town: Unpublished.

Operation Iraqi Freedom. 2003. President Bush addresses the nation, The White House. Retrieved 16 November 2005 from World Wide Web: www.whitehouse.gov/news/releases/2003/03/20030319-17.html.

Pearce, F. 2003. NewScientist.com: Heat will soar as haze fades. June 7. [Online]. Available: http://www.newscientist.com/channel/earth/climate-change/mg17823980.800. [Accessed 5 April 2005].

Phillips, K. 2004. Bush family values: war, wealth, oil. Latimes.com. February. [Newspaper, selected stories online]. Retrieved 17 November 2005 from World Wide Web: www.latimes.com.

Ponting, C. 2000. *World History. A New Perspective*. London: Random House, Pimlico.

Radford, T. 2005. Stark warning to the world. *Mail & Guardian*: 28, April 1–7.

Reuters. 2005. Netscape News: 'Dangerous' global warming possible by 2026 – WWF. January 29. [Online]. Available: http://channels.netscape.com/ns/news/story.jsp?id=20050129190300002586895&dt=20050129190300&w=RTR. [Accessed February 2005].

Richter, A., Burrows, J.P., Nüß, H., Granier, C. and Neimeier, U. 2005. Increase in tropospheric nitrogen dioxide over China observed from space. *Nature*. 437: 129–132.

Roberts, P. 2004. *The End of Oil. The Decline of the Petroleum Economy and the Rise of a New Energy Order*. London: Bloomsbury.

Sparks, D. and Mwakasonda, S. 2004. Energy and the environment. In Energy for Sustainable Development – South African: Profile Phase 1 Final Report. Energy Research Centre, University of Cape Town. Unpublished.

Townsend, M. and Harris, P. 2004. *Guardian Unlimited*: Now the Pentagon tells Bush: climate change will destroy us. February 22. [Online]. Available: http://observer.guardian.co.uk/international/story/0,6903,1153513,00.html. [Accessed 5 April 2005].

Turpie, J.K., Heydenrych, B.J. and Lamberth, S.J. 2003. Economic value of terrestrial and marine biodiversity in the Cape Floristic Region: implications for defining effective and socially optimal conservation strategies. *Biological Conservation*. 112 (2003): 233–251.

United Nations Framework Convention on Climate Change. 2000. South Africa: Initial national communication under the United Nations Framework Convention on Climate Change. [Online]. Available: http://unfccc.int/2860.php. [Accessed November 2004]. [Digital copy in possession of author].

United Nations Framework Convention on Climate Change. No date. Essential background: feeling the heat. [Online]. Available: http://unfccc.int/essential_background/items/2877.php. [Accessed 22 May 2005].

Wilson, E.O. 2002. *The Future of Life*. London: Abacus.

Winkler, H. 2004. Energy demand. In Energy for Sustainable Development – South African: Profile Phase 1 Final Report. Energy Research Centre, University of Cape Town. Unpublished.

World Resources Institute. 2005. CAIT Climate Analysis Indicators Tool: Total GHG emissions in 2000. [Online]. Available: http://cait.wri.org/cait.php?page=yearly. [Accessed 6 April 2005].